Reconstruction and Reform

A HISTORY OF US

The picture on the cover is titled *George W. Hoag's Record Wheat Harvest*. Hoag had a farm near Sacramento, California, where 50 machinists worked at seven forges building a 35-foot-long red-painted monster thresher. Hoag named it *Monitor*, and, with a crew of 56 men and 96 horses and mules, *Monitor* set a world threshing and bagging record in 1874. Then Hoag hired artist Andrew Putnam Hill to record the event. That's Hoag right there in the center in the black buggy.

Oxford University Press

OXFORD
A HISTORY OF
US

BOOK SEVEN

Reconstruction and Reform

Joy Hakim

Oxford University Press
New York

Oxford University Press

Oxford   New York   Toronto
Delhi   Bombay   Calcutta   Madras   Karachi
Kuala Lumpur   Singapore   Hong Kong   Tokyo
Nairobi   Dar es Salaam   Cape Town
Melbourne   Auckland   Madrid

and associated companies in
Berlin   Ibadan

Copyright © 1994 by Joy Hakim

Designer: Mervyn E. Clay
Maps copyright © 1994 by Wendy Frost and Elspeth Leacock
Produced by American Historical Publications

Published by Oxford University Press, Inc.
200 Madison Avenue, New York, New York 10016
Oxford is a registered trademark of Oxford University Press

Library of Congress Cataloging-in-Publication Data
Hakim, Joy.
Reconstruction and reform / Joy Hakim.
p.   cm.—(A history of US: bk. 7)
Includes bibliographical references and index.
ISBN 0-19-507757-1 (lib. ed.)—ISBN 0-19-507758-X (pbk.)
1. United States—History—1865–1898—Juvenile literature.
2. Reconstruction—Juvenile literature. 3. United States—Social conditions—1865–1918—Juvenile literature.
[1. United States—History—1865–1898. 2. Reconstruction.] I. Title. II. Series: Hakim, Joy. History of US; 7.
E178.3.H22 1994 vol. 7
[E661]
973.8—dc20   93-26251
CIP
AC

1 3 5 7 9 8 6 4 2
Printed in the United States of America
on acid-free paper

The stanzas on pages 5 and 70 are excerpted from *Western Star* by Stephen Vincent Benét. Copyright © 1943 by Rosemary and Stephen Vincent Benét. Copyright renewed © 1971 by Thomas C. Benét, Stephanie B. Mahin, and Rachel Benét Lewis. The stanzas on pages 90 ("Crazy Horse") and 102 ("Barnum") are excerpted from *A Book of Americans* by Rosemary and Stephen Vincent Benét. Copyright © 1933 by Rosemary and Stephen Vincent Benét. Copyright renewed © 1961 by Rosemary Carr Benét. All reprinted by permission of Brandt & Brandt Literary Agents, Inc. The passage on page 60 is excerpted from *China Men* by Maxine Hong Kingston, copyright © 1980 by Maxine Hong Kingston. Reprinted by permission of Alfred A. Knopf, Inc. The stanzas on page 180 are from "Booker T. and W. E. B.," excerpted from *Poem Counterpoem* by Dudley Randall. and Margaret Danner. Copyright © 1966 by Dudley Randall. Reprinted by permission of Broadside Press, Detroit, Michigan.

Americans are always moving on.
It's an old Spanish custom gone astray,
A sort of English fever, I believe,
Or just a mere desire to take French leave,
I couldn't say. I couldn't really say.
But, when the whistle blows, they go away.
Sometimes there never was a whistle blown,
But they don't care, for they can blow their own
Whistles of willow-stick and rabbit bone,
Quail-calling through the rain
A dozen tunes but only one refrain,
"We don't know where we're going,
       but we're on our way!"
                    —STEPHEN VINCENT BENÉT,
                         WESTERN STAR

Always do right. This will gratify some people and astonish the rest.

—MARK TWAIN,
ADVENTURES OF HUCKLEBERRY FINN

The most necessary task of civilization is to teach people how to think. It should be the primary purpose of our public schools. The mind of a child is naturally active, it develops through exercise. Give a child plenty of exercise, for body and brain. The trouble with our way of educating is that it does not give elasticity to the mind. It casts the brain into a mold. It insists that the child must accept. It does not encourage original thought or reasoning, and it lays more stress on memory than observation.

—THOMAS EDISON

Didn't my Lord deliver Daniel,
D'liver Daniel, d'liver Daniel?
Didn't my Lord d'liver Daniel,
And why not every man?

He deliver'd Daniel from the lion's den,
Jonah from the belly of the whale,
And the Hebrew children from
the fiery furnace,
And why not every man?

—NEGRO SPIRITUAL

Contents

A sharecropping family living in the South in the years after Emancipation.

PREFACE
Are We Equal?
Are We Kidding?

With the Civil War over, a Union soldier and his wife pose proudly for their picture. Blacks fought and worked for their country. What did they deserve in return?

The United States began with an idea —a never-tried-before idea—the idea that all people are created equal.

What do you think of that idea? Are we all equal?

Of course not.

Some people have voices good enough to be opera stars. Some of us can't even sing in tune. We certainly don't have equal musical ability.

Some people can easily add up big lists of numbers in their heads. Others of us struggle with pencil and paper.

Can you see where we are going with this? We are not equal when it comes to ability. People are different.

So what is it that we Americans mean when we insist that *all men and women are created equal*?

We mean that all of us have certain rights. We hold them equally. No one has more of those rights than anyone else. In the Declaration of Independence, Thomas Jefferson said they are the rights to *life, liberty and the pursuit of happiness*.

Life? Are you saying, "Of course we have the right to be alive!"

Not so fast. People in this country are more fortunate than you may realize. In recent years, in some nations, dictators have killed millions of their own citizens. In the United States no ruler or governing body can take your life unless a jury convicts you of a terrible crime.

Some people think the right-to-life issue has not been decided in America. They ask: What about capital punishment? What about abortion?

Capital punishment is the death sentence for a crime. Many nations—and many of our states—have abolished capital punishment. They keep criminals in jail but do not kill them.

9

We hold these Truths to be self-evident, that all Men are with certain unalienable Rights, that among these are

Abortion is the expulsion of a fetus (an unborn child) from a pregnant woman's womb, before the fetus is ready to live outside.

The liberty issue was settled in 1865. The Civil War decided it. *No more slavery.* It was finished. Finally finished. Or was it? Are you really free if you don't have an education and can't get a job? What is liberty?

And how about the pursuit of happiness? How does that fit with equality? Does it mean that we should all be free to pursue any honest path in life that we wish—without regard to our skin color or religion or ethnic background? Most people think so.

In some nations, all the citizens are of the same ethnic background. That means they have common ancestors and look alike. The United States sprang from a different idea. We don't exclude anyone because of race or national origin. Sometimes that creates problems, but it also makes this country interesting. And guess what? When people of different backgrounds get together, they usually find they have much in common.

The United States is still an experiment. Our Founding Fathers gave us remarkable goals. For more than 200 years we have been trying to build a nation where peoples of all colors and religions and abilities are welcome and treated equally. No nation has ever done that. It isn't easy.

This photograph was taken in Richmond, Virginia, after the war was over, and the store owner, the man in the long apron, is celebrating something. Can you figure out what the occasion is?

created equal, that they are endowed by their Creator Life, Liberty and the Pursuit of Happiness . . .

A Water Bird and a Sparrow: Or, How Would Abraham Lincoln Reconstruct?

It was February 1865, and Abraham Lincoln was standing on the deck of the steamboat *River Queen*, moored off the Virginia coast near Hampton Roads (the water highway that took the first English settlers to the James River and Jamestown). The lean, big-boned president must have looked like a long-legged water bird next to his sparrowlike guest: frail, wizened Alexander Stephens. Stephens was vice president of the Confederate States of America.

The Confederacy was made up of 11 southern states that had decided to separate themselves from the rest of the United States. Those states were attempting to create a nation of their own—a slave nation. The North wasn't letting them do it. The union of states can't be broken any time a state decides it doesn't agree with the others, said the Northerners. Besides, slavery was wrong, they insisted. Those two issues, slavery and states' rights, had led to war—civil war—the worst war in our nation's history. These two men—Lincoln and Stephens—were war leaders, and, presumably, enemies. And so this meeting was unusual. But they were both men of good will; they were meeting to talk about the future. They were meeting to talk about peace.

It was now clear that the Civil War would soon be over. It was just a matter of time before the Confederate armies must surrender. Lincoln hoped to speed up that surrender. He hoped to

Abraham Lincoln greets Confederate vice president Alexander Stephens and his delegation with open arms and a sly smile. He knows that the end of the war is only a matter of time.

convince Stephens to help end the war now. But the Confederacy's vice president still insisted on independence for the South. (The war would continue for two more months.)

President Lincoln meant to reunite North and South. After four years of war he knew that wouldn't be easy. Lincoln wanted to reestablish harmony. He wanted to erase the hatreds. He wanted to reconstruct the nation into a harmonious whole. But how did he intend to do it? What plans did he have for the South's slaves, who were soon to be free? What did he tell Alexander Stephens? If there are written records of their conversation, no one knows of them.

Later, when others were leading the country, people asked, "What would Lincoln have done to reconstruct the nation?" For a powerful clue, all they had to do was read the Declaration of Independence.

All men are created equal—it couldn't be clearer. The Declaration was "the great fundamental principle upon which our free institutions rest," said Abraham Lincoln.

And it wasn't just for some Americans. Lincoln said the Declaration was not merely "the white man's charter of freedom." All Americans, he said, are "entitled to… the right to life, liberty, and the pursuit of happiness."

Unfortunately there were some citizens who just didn't seem to understand. Maybe they had never read the Declaration of Independence. Or maybe they hadn't thought about what it says.

11

1 Reconstruction Means Rebuilding

In 1866, Mississippi spent a fifth of state income on artificial arms and legs for war veterans.

In the North one hears the war mentioned, in social conversation, once a month; sometimes as often as once a week; but as a distinct subject for talk, it has long ago been relieved of duty....The case is very different in the South. There, every man you meet was in the war, and every lady you meet saw the war. The war is the great chief topic of conversation. The interest in it is vivid and constant; the interest in other topics is fleeting....In the South, the war is what A.D. is elsewhere: they date from it.

—MARK TWAIN,
LIFE ON THE MISSISSIPPI

The Civil War was over, and all across the land mothers and fathers buried their sons, wept, and tried to forgive the enemy now that they were pledging allegiance to the same flag. Most people seemed to understand that the country had to be made whole again. Its wounds needed to be bandaged.

President Lincoln had been determined to use kindness in bringing the South back into the Union. Actually, Lincoln said the South had never left the Union. Some southern people had rebelled—that was what had happened, Lincoln said. It was like a family fight. They were still part of the family. Lincoln wanted to make it as easy as possible for the nation to reunite. Others felt differently.

Some Northerners were very angry. After all, it was the South that had started the war. It had been more terrible than anyone could have imagined. How should the North treat its former enemy? Should it be punished? Some thought the Rebel leaders should be hanged.

Did you ever lose a fight? Were you embarrassed and angry? White Southerners were angry, confused, hurt, and miserable. You can understand that. Their lovely,

A poster for a show of left-handed penmanship by veterans who had once been right-handed and now had to learn to write all over again.

"War is hell," said General Sherman. This is the railroad depot in Charleston, South Carolina, after Sherman's troops passed by.

elegant, aristocratic South was in ruins. Their sons were dead. Everything they had fought for seemed gone. (*Gone with the wind*, said one southern woman in a famous book.)

A visitor to Charleston, South Carolina, wrote of "vacant houses, of widowed women, of rotting wharves, of deserted warehouses, of weed-wild gardens, of miles of grass-grown streets." Most of the South's cities were in the same shape. And the countryside? "We had no cattle, hogs, sheep, or horses or anything else," a Virginian wrote. "The barns were all burned, chimneys standing without houses and houses standing without roofs, or doors, or windows." Across the South

Immigrants Are Changing Things

It wasn't just war and Reconstruction that were happening in the 1860s. People were pouring into the country. They were emigrating from Ireland, China, Scandinavia, England, Greece, and other nations, and they were bringing new ideas and new skills. After the Civil War—in spite of the 620,000 war deaths—there were more people in the country than when the war began.

Many of those people settled in the north, but others headed west. In one western region there were so many people from Germany that the Native Americans began speaking German. Some areas had Swedish speakers, some had Russians, others had Danes. Some of the territories became as *polyglot* ("many-tongued") as New York City.

everything seemed collapsed and disordered. There was no government, no courts, no post offices, no sheriffs, no police. Guerrilla bands looted at will.

A generation of white Southern men was dead. Those who came home brought wounds with them. In 1866, the year after war's end, Mississippi spent one fifth of its revenues on artificial arms and legs for returning veterans.

Southern whites had to blame someone for their misery, and people don't like to blame themselves, so the former Rebels blamed Northerners. They said everything that went wrong after the war was the Northerners' fault. And as for the Civil War itself, all they had tried to do, they said, was form their own nation. How could they forgive the North for stopping them?

What of the four million black Southerners who were now freedmen and freedwomen? What were they to do now? For many, freedom meant

This family was free. But without land, without law and order, without civil rights backed by guarantees, what did "freedom" mean?

going somewhere—anywhere. But where were they to go? What were they to do? Should they be paid for all their years of past work? There were rumors that each former slave would get 40 free acres of land and a mule to work it. Would that happen?

Most of the ex-slaves couldn't read or write. They wanted to learn. Who would their teachers be? Many had no idea what freedom really meant. Some thought it meant they would never have to work again.

Someone needed to do some organizing. Someone needed to maintain law and order. Help was needed.

The time in the South after the Civil War, when people attempted to reorganize and remake the region—without slavery—is called *Reconstruction*.

How did it go? With a whole lot of confusion. It was the most promising, despairing, noble, awful, idealistic, reactionary, hopeful, hopeless time in all of American history. It didn't end up very well.

2 Who Was Andrew Johnson?

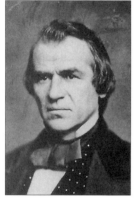

"I love my country," Andrew Johnson said. "Every public act of my life testifies that is so."

It was an actor's bullet that gave the country a new president. Now everyone was asking what kind of a man he was. People didn't know quite what to expect of President Andrew Johnson.

He was almost exactly the same age as Abraham Lincoln, and, like the old rail splitter, he'd been a poor boy who had made his own way in life. Johnson had once been a tailor, but when he got up at a political meeting and began speaking, he found he had a talent. He could captivate and hold an audience.

Andy Johnson didn't have much education, but he was smart and he soon became prosperous. His wife helped smooth his rough edges and taught him some book learning, too. He was a Democrat and a slave owner.

In his home state, Tennessee, he became governor, congressman, and senator. He was in the Senate when the Southern states, including Tennessee, seceded. He stuck with the Union. When Union forces captured Tennessee, President Lincoln made him military governor. He held that post during the war. He was often in danger. But Andrew Johnson was courageous. He was a good man for the job.

Johnson as a cartoon parrot, always squawking about the Constitution.

Lincoln was sometimes called the rail splitter because he had earned money splitting logs for railroad ties.

On Depot Street in Greeneville, Tennessee, you can visit Johnson's tailor shop (above), as well as his house and grave.

Buying a Piece of Hot Ice

In 1867, Secretary of State William Seward bought a big, important piece of land for the United States. Some people thought it worthless and overpriced. They called it "a large lump of ice." Seward spent $7.2 million, or about 2 cents an acre, for the purchase. Can you guess what he bought?

He bought Alaska from Russia. Some Americans wanted Russia to keep it. They called the deal "Seward's folly." (What is a *folly*?) Seward's icebox (it was called that, too) turned out to be one of the world's great real-estate deals, like the buying of Manhattan island, or the Louisiana Purchase. (What was the Louisiana Purchase, and when did it happen?) In 1896, when gold was discovered in the Klondike, people decided that Seward was smart after all.

Cartoons like this one (left), of a politician reduced to soliciting votes from polar bears, were everywhere when Seward (above) was buying Alaska.

Even though Johnson was a Democrat, and Lincoln a Republican, Abraham Lincoln asked Andrew Johnson to be vice president.

Now the awful war was over. It was time for healing. Most people were encouraged. Johnson seemed the perfect person to bring South and North together again. After all, Johnson was a Southerner who had had the courage to stay with the Union. Both Democrats and Republicans supported him. Perhaps it was all for the best.

But those who knew Johnson weren't so sure. Yes, he had courage, no doubt about that. But he was also stubborn. Mulishly stubborn. He didn't ask for advice, or listen when it was given. Lincoln asked questions, listened, and changed his mind when he thought it needed changing. He knew how to compromise. Andy Johnson was uncompromising. He was like a wall: rigid. You know how walls act. They don't bend, but with enough force, you can break them.

Well, the wall was about to be charged.

Andrew Johnson, tailor, stitches the tattered Union suit back together in 1865.

Back in 1852, when Andrew Johnson ran for governor, the Know-Nothing Party was whipping up anti-Catholic feeling in Tennessee, where most people were Baptists or Methodists. Johnson would have none of that. He said American Catholics were American citizens—and won the election.

3 Presidential Reconstruction

Two newly freed Louisiana slave children wore brand-new clothes to have their picture taken after emancipation was proclaimed in 1863.

During the first two years of Reconstruction, President Andrew Johnson was in control. That time is called *presidential Reconstruction.*

In the beginning, things seemed to go well.

Congress had created a Freedmen's Bureau even before the war ended. It was to help the newly freed blacks. They needed food, clothing, and shelter. Some Northerners went south to help. Many of them were teachers.

The Freedmen's Bureau began opening schools. Slaves had been starved for learning. In the years of the Confederacy, every Southern state except Tennessee had laws making it a crime to teach slaves to read and write. Now, as free people, they were thirsty for knowledge. In Mississippi, when a Freedmen's Bureau agent told a group of 3,000 they were to have schools, he reported that "their joy knew no bounds. They fairly jumped and shouted in gladness." When schools opened, parents often sat in classrooms with children. As soon as they could read and write, the new learners taught others.

Many of their teachers were white missionaries sent by northern churches. Others, like Mary Peake (who founded the first school for blacks in Hampton, Virginia), were educated northern blacks.

But it's hard to learn if you're hungry, and southern farms were in terrible shape. In 1865 the wheat crop failed. It didn't do much better the next year.

The Freedmen's Bureau kept most people from starving. Clothing was distributed. In some places more of its help went to whites than blacks. There was nothing wrong with that. Anyone who needed help

Many teachers who went south to instruct former slaves felt as if they were in a foreign land. One teacher in Georgia said, "Our work is just as much missionary work as if we were in India or China."

Animosity is hatred or ill feeling.

It seemed like it took a long time for freedom to come. Everything just kept on like it was. We heard that lots of slaves was getting land and some mules to set up for themselves. I never knowed any what got land or mules nor nothing.

—MILLIE FREEMAN, FORMER SLAVE

was meant to have it.

And people helped each other. A former house slave found a job and, each week, brought five dollars to his old mistress. Many white Southerners helped the freedmen and freedwomen.

Northern soldiers kept order. A citizen of Lynchburg, Virginia, said, "A more gentlemanly and humane set of officers, and I may add, of soldiers, never occupied an enemy's country." But not everyone agreed. Just looking at those blue uniforms upset many Southerners. And some whites couldn't accept the idea of a society where people were equal.

Thousands of Southerners left the country for Mexico and South America. Some Confederates went to Brazil, where there was still slavery (although abolition soon came to that nation, too). General Lee was not pleased to see Southerners leave the United States. "Virginia has need of all her sons and can ill afford to spare you," he wrote one of them. To others he said, "Abandon all these local animosities and make your sons Americans."

Most Americans put the war behind them. Northerners and Southerners were soon visiting each other again. Young southern men went north to enroll in the military academies at West Point and Annapolis. And some North–South romances blossomed (although Jefferson Davis wouldn't let his daughter marry a Yankee).

A few Southerners attempted a new kind of cooperation between the races; their experiments in racial harmony wouldn't be tried again for a hundred years. But, mostly, white people in the South still didn't understand. They were willing to be good United States citizens, except—and it was a big exception—except when

Uncle Sam says he will take off the "dunce cap" of martial law as soon as the naughty southern child is good.

In this drawing a Union officer stands between angry whites and angry blacks. The artist was trying to show that soldiers were needed to keep peace in the former Confederacy. Congress did send northern troops south to maintain order and protect the rights of the freed persons. Soldier rule is called *martial law*. (See chapter 5 for more about this.)

it came to treating their black fellow citizens fairly. And, once again, they turned to the wrong leaders.

Right after the war (in 1865 and 1866) most of the same old southern leaders were in charge, and every southern state passed laws that discriminated against blacks. The laws were called *black codes*. They made blacks practically slaves again. The codes gave whites almost unlimited power. No southern state would establish public schools for blacks.

There were outbreaks of violence against blacks in many areas of the South. Race riots erupted in Memphis and New Orleans. Whites were attacking blacks. General Philip Sheridan, who was at New Orleans, called the riot there a massacre. Thirty-four blacks and three whites (who stood with them) were killed. More than 100 people were injured. Some whites put masks over their faces and began acting like hoodlums. They went around burning black churches and schools; they terrorized and killed blacks. These were grownups. They were members of a newly formed hate organization, the Ku Klux Klan, and they didn't have the courage to show their faces.

The state governments did not bring them to justice.

There had been four years of warfare over slavery and states' rights and it still wasn't over. Now the fight had switched from the battlefields to the halls of Congress.

This cartoon showed Johnson as a callous Roman emperor watching the massacre of innocents in the riots at Memphis and New Orleans.

Ku Klux Klan members' white masks had red-braid holes for eyes, nose, and mouth. They terrorized black people, sometimes pretending to be the ghosts of Confederate soldiers.

The southern states sent representatives to Congress. Many were former leaders of the Confederacy: generals and colonels. Even Alexander Stephens, who, as you know, had been vice president of the Confederacy, was sent to Congress. Northerners were outraged. After many wars, defeated leaders are tried and sometimes even executed. Before the war, many southern officers had been in the U.S. Army. They had sworn allegiance to the United States and then had broken that pledge and fought against the nation. The North had not asked for revenge. General Grant had paroled the Rebel soldiers and officers. But should they be rewarded and made congressmen?

And what of those other things? "Black codes" that bound black workers to labor contracts and gave them no legal rights? That sounded like slavery. Race riots? Murders? That wasn't American justice. "Why did we fight a war?" people in the North asked.

President Johnson urged the southern states to protect the freedmen and freedwomen's rights. But he didn't do anything to see that they were protected. There was something else disturbing. He was being nasty to Southerners who had supported the Union (as he had). He seemed to be taking sides at a time when he should have tried to be president of all the people.

What was going on?

Something very important was going on. The war had been fought over issues that would determine what our nation was to stand for. Those issues had not been settled.

What were they?

They Were Unusually Bright Today

Even before the war ended, when Union troops occupied parts of the South, some Northerners came to teach the former slaves. Charlotte Forten came from Philadelphia to Port Royal, South Carolina. Forten's grandfather was James Forten, a prominent abolitionist (if you don't remember him, see Book 3 of *A History of US*). Charlotte had just finished school herself, but she was soon busy teaching. Here is the entry in her journal for a December day:

A truly wintry day. I have not had half as many scholars as usual. It was too cold for my "babies" to venture out. But altogether we had nearly a hundred. They were unusually bright today, and sang with the greatest spirit....After school the children went into a little cabin near, where they had kindled a fire, and had a grand "shout."

She met some interesting people in Port Royal; Harriet Tubman was one of them. After she returned north, Charlotte Forten married the Reverend Francis J. Grimké, whose two amazing aunts were well-known abolitionists. (And if you don't know about them, look in an encyclopedia, or in Book 5.)

4 Slavery and States' Rights

Amendment 13

Section 1. Neither slavery nor involuntary servitude, except as a punishment for crime whereof the party shall have been duly convicted, shall exist within the United States, or any place subject to their jurisdiction.

Section 2. Congress shall have power to enforce this article by appropriate legislation.

The easiest way to suppress black voters was by violence and terror. This illustration was called *One Vote Less*.

The 13th Amendment was ratified on December 6, 1865. That did it. It ended slavery. The Emancipation Proclamation was now the law of the land.

But some thinking people were already asking themselves, "Is being free of slavery enough?" If you are free and can't vote, are you really free? If you are free but laws say you can't quit your job or leave your plantation—as the black codes said—then are you really free?

Those thinking people started talking about the whole meaning of America. The United States was still an experiment. It was still the only people's government around. Those thinkers went right back to the Declaration of Independence.

We hold these truths to be self-evident, that all men are created equal, that they are endowed by their Creator with certain unalienable Rights, that among these are Life, Liberty and the pursuit of Happiness.

Southern whites used all kinds of rules and tricks to keep blacks from voting.

Unalienable, inalienable
—which is correct? Both: Jefferson used *un*alienable; *in*alienable is usually used today.

21

It was war between President Andrew Johnson and the Radical Republicans. Once the Republicans realized they could pass laws over Johnson's veto, they took control of Reconstruction.

Radical means extreme. It means going as far as you can go with an idea or a change.

Nullify means to negate or make into nothing. In this case the Civil Rights Act was meant to make the black codes ineffective; to stop the anti-black laws from working.

A **veto** is a *no* vote. The president's veto usually means that a law will not be passed.

Unalienable rights. That means rights that belong to each of us—not to the government. No government has any right meddling with our rights to life, liberty and the pursuit of happiness, says the Declaration.

Before 1865, the southern states had pretended that blacks didn't have those same rights. They had pretended that blacks were different. They had acted as if unalienable rights were only for whites.

The war had been fought to end slavery. Slavery was ended. But the black codes were there to do the same old thing: to keep blacks from their unalienable rights.

A group of leaders in Congress said, "Hold on." And then they said, "Stop." They wrote laws to protect the civil rights of blacks. Those congressional leaders were a small group within the Republican Party. They were called Radical Republicans. In 1866 the Radical Republicans got Congress to pass a Civil Rights Act. It was designed to nullify the black codes.

President Johnson vetoed the Civil Rights law. Johnson was a former slave owner. Although he no longer believed in slavery, he did not believe in equality. Johnson also thought it was up to the states, not the central government, to protect individual rights. (That was the *states' rights* argument.)

Congress passed the Civil Rights law over the president's veto. (Study the Constitution to find out how that is possible.) It was the first time in American history that an important piece of legislation was passed over a president's veto. As you know, Andrew Johnson was stubborn. But so were the Radical Republicans. The battle between them now grew fierce.

The next thing the Radicals did was to write the 14th Amendment. It is a long but *very important* amendment. Here is a key part of it:

> *No State shall deprive any person of life, liberty, or property, without due process of law; nor deny to any person within its jurisdiction the equal protection of the laws.*

In 1867, the black codes that had been introduced after the Civil War were abolished and former slaves, like these men in Macon, Georgia, were able to register to vote.

The jury in this southern courtroom was mixed—but the races kept apart, blacks with blacks, whites with whites.

That amendment says no state can take away any person's rights. We are all entitled to the safeguards provided by the Constitution—even against a state!

Back in 1787, the Constitution makers had worried about protecting citizens from abuses by Congress, or the president, or the federal government—but not from abuses by their states. Each state was expected to protect its own citizens. But suppose they didn't? This amendment was saying that the U.S. government would protect all its citizens, even against the states. The United States government was to be superior to any state.

Whew! This was a powerful amendment. The South had fought for states' rights. Many Southerners thought each state should be free to make its own decisions. If a state wanted an aristocratic society with layers of privilege and unfairness—well, if that was what the majority of its people really wanted, why shouldn't they have it?

The Constitution makers had thought about that. They said majorities were sometimes tyrannical. They worried about the rights of minorities. The Bill of Rights was to protect minorities even from majority will.

Were we to stick with the goals of the Founders? Were we to respect the inalienable rights of all citizens?

Should we be a nation with a constitution that guarantees fairness to all? If so, then if people vote for unfair state laws, those laws need to be made unconstitutional. And that is just what the 14th Amendment does.

It was a big step. Andrew Johnson didn't like it a bit. It took power from the states. It gave power to the Supreme Court. Johnson didn't like the 14th Amendment and he hated the man who was the force behind it. (For details, see chapter 6.)

Free at Last...

Former slaves rejoice in 1866 at the passing of the Civil Rights Act.

When black men, women, and children learned they were free, there was singing and shouting "an' carryin' on." Charlotte Brown, who lived in Wood's Crossing, Virginia, remembered that. Then, she said, a few days later, on Sunday morning, "we was all sitting roun' restin' and tryin' to think what freedom meant an' everybody was quiet and peaceful." They knew freedom meant being able to move wherever they wanted, to find work wherever they could, to choose new names, to plant their own gardens, to search for family members, to practice their own religion, and to go to school and learn.

We belong to ourselves," they shouted out, and after years of slavery, that was enough for some. But others realized that freedom was more than that. Sometimes they were surprised to find that freedom means choices and responsibilities. Just what is freedom? What does it mean to you?

5 Congressional Reconstruction

Most Southerners hated carpetbaggers, as this cartoon shows, and thought they wanted only to profit from the South's problems.

In those days a traveling bag was sometimes made of carpet material. Some Southerners said the carpetbaggers were people who threw a few things in a suitcase and came South only to make money for themselves. Mostly, that was a myth.

Southerners who cooperated with the North were called *scalawags*. Some scalawags wanted power and influence. But some, like James Longstreet (the man General Lee affectionately called "my old war horse"), just wanted to forget the war and do business in a united nation.

Congress decided to send soldiers south to guarantee freedom to the former slaves. That happened in 1867. The soldiers stayed for about 10 years. That period is called *congressional Reconstruction*. (You may also hear it called *military Reconstruction*.) During that time, many Northerners went South. They went to teach, to help with aid programs, to help the state governments get going again, and sometimes to make money for themselves.

All those Yankees were known as *carpetbaggers*. Many Southerners found it hard to put up with Northerners in their midst, especially Yankees who were telling them how to behave. Most white Southerners who had been Confederates hated the carpetbaggers. The carpetbaggers reminded them of the war and their losses.

Some Northerners did take advantage of the South. Often they were businessmen who didn't even bother to visit the region. Northern lumbermen looked greedily at southern forests. In 1876 they got Congress to sell southern land cheaply. One congressman took 110,000 acres of Louisiana land for himself. A Michigan firm got 700,000 acres of pine land. Another northern firm bought more than a million acres at 45 cents an acre. What they did has been called the most reckless destruction of forest land in history.

But most of the Northerners who went South went to help. They wanted to see blacks treated fairly. Many had great courage. Many worked in the Freedmen's Bureau. For five years that bureau set up schools, distributed food, and helped the newly freed citizens with their problems.

Congress passed a Reconstruction Act. Naturally, President Johnson vetoed it. Enough votes were gathered in Congress to get the act passed over his veto. The act said that to become part of the Union again, each southern state had to write a new constitution that was true to the U.S. Constitution. The act also said that all males over 21 could vote, except for convicted criminals and those who had participated "in the rebellion." That meant that former Confederate soldiers could not vote, but black men could. As you might guess, that made some whites angry and bitter.

Northern soldiers made sure that black men were able to vote. It was amazing. Men who had been slaves a few years earlier were lining up at the polls. Many were illiterate (they couldn't read or write), but about one fifth of the South's white population was illiterate, too. Being illiterate doesn't mean being stupid. The new voters did exactly what James Madison expected them to do. They voted for what they believed to be their own interests.

Madison had thought that made sense. In a large democratic nation, he said, all the special interests need to be heard. They would balance each other. Before the war only the interests of the planters had been represented. Now many blacks were being elected to office. And some poor whites, too.

Mississippians Blanche K. Bruce and Hiram R. Revels became U. S. senators. Both were college men. Revels

The three pillars of the Democratic Party—Irish immigrant, Confederate veteran, and Wall Street financier—trampling the innocent black man.

Some planters—with crops like "Reconstruction tobacco"—hoped for quick profits, but in 1866 and 1867 crops failed and recovery was slow.

Above, the senators from Mississippi, Blanche Bruce (left) and Hiram Revels (right).The name "Blanche" is from the French word for "white."

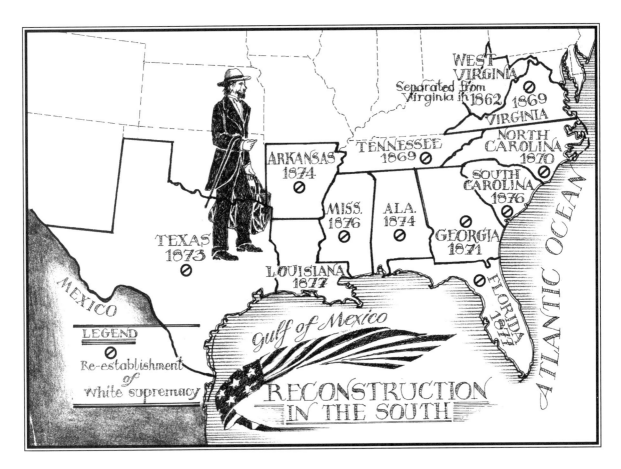

LEGEND

⊘ Re-establishment of white supremacy

RECONSTRUCTION IN THE SOUTH

Sixteen blacks served in the U.S. Congress during Reconstruction; more than 600 were elected to state legislatures; and hundreds more held local offices like sheriff or justice of the peace. Blacks and whites served together on juries, on school boards, and on city councils. They ensured fairness in ways that had not been done before (and would not be done again until the second half of the 20th century).

took Jefferson Davis's old seat in the Senate. The day he was sworn in the Senate galleries were packed, and everyone stood as he walked down the aisle. Some observers burst into cheers to see a black as a senator. "Never since the birth of the republic," said an editorial in the *Philadelphia Press*, "has such an audience been assembled under one single roof. It embraced the greatest and the least American citizens." That was an exaggeration, but those who were there knew it was a historic moment.

Bruce, a wealthy farmer and landowner, had attended Oberlin College. In the Senate he worked for equal rights for all. Few people, then, worried about rights for Indians and Asians. But Bruce did. He was an eloquent, dignified senator who worked to help the growing railroads and to improve navigation on the Mississippi River. At a time when there was much corruption in government and business, Blanche Bruce was known for his integrity.

Still, it was difficult for some white people to accept what was happening. Blacks in Congress! Blacks in the state legislatures! Some whites really believed the racist myths. They didn't believe blacks could think as well as whites. They didn't believe women could think as well as men. In the North most white men wouldn't take a chance (and there weren't many black voters), so only a few blacks were elected in northern states. And no women were elected at all—north or south—because women still couldn't vote.

This cartoonist drew Andrew Johnson as Iago, the villain and betrayer of the black man in William Shakespeare's play *Othello*.

From Railroad Terminus to Capital City

Atlanta was one city that tried not to live in the past; it described itself as the "New South." Maybe that was because it was a young city, founded only 23 years before the Civil War began. At first it was just called "Terminus," because it was where the Western and Atlantic Railroad terminated (which means ended). But when it started to grow—the railroad made that happen—it got a proper name.

Georgia's capital was at Milledgeville, but after the war, Major General John Pope, commander of the federal occupation army, put his headquarters in Atlanta. So when a convention was called to write a new constitution, it met in Atlanta. Then members of the Atlanta city council offered the city's opera house as a rent-free home for the legislature, if the capital was moved to Atlanta. It was.

6 Thaddeus Stevens: Radical

"Who is the United States?" said Stevens. "Not the judiciary, not the president, but the sovereign power of the people."

This chapter is about President Andrew Johnson's enemy.

When he was a boy Thaddeus Stevens had three big problems: he was very poor, his father was an alcoholic who deserted the family, and he was handicapped. Thaddeus had a deformed leg and foot—a clubfoot—that made him limp badly. The leg made him self-conscious, and some boys and girls made fun of him. That didn't help.

But Thad had some things going for him: he had an amazing mind and he was fiercely honest. He was handsome, too. He could have been a model for a Greek sculptor. His face had strong, regular features and deep-set hazel eyes. His curly hair was the color of chestnuts.

Thaddeus Stevens grew up in Vermont, where people say what they think and don't waste words. When he was 24, in September 1816, he moved to a pretty little Pennsylvania town. You've heard of it. It was called Gettysburg.

He was soon the best lawyer in Gettysburg. And then the best in Pennsylvania. The richest, too. He owned an iron foundry. In addition, he was a shrewd investor: he took his money and put it in other businesses and made more money. His Gettysburg neighbors thought well of him; they sent him to Congress, and kept him there for many years.

White Southerners would soon call him the vilest of all Yankees. Blacks would see him as a hero, and rank him close to the angels. Those who knew him best either admired or hated him—you couldn't have halfway feelings about Thad Stevens.

He didn't try to make himself liked. He went around with a frowning, stern look on his face. He thought most people were selfish and

Thaddeus Stevens's iron foundry was destroyed by Robert E. Lee's troops at the battle of Gettysburg.

many were evil. He never married. An illness left him bald in his thirties, so he wore a dark wig. The wig didn't fit well.

In those days most people went to church every Sunday. Not Thaddeus Stevens, though he had been raised a Baptist. People whispered about him, because he wouldn't do as others did, but when he heard that two young men wanted to study to be ministers, and didn't have the money for tuition, he quietly paid for their schooling.

Once he was riding his horse home when he saw a crowd in front of a farm. He stopped and learned that a widow's farm was being auctioned. The widow had no money to pay her bills. So he bought the farm, gave it back to the widow, got on his horse, and quickly rode off.

Stevens did what he wanted, said what he wanted, and didn't seem to care a bit what others thought. He was unwaveringly honest and couldn't be bribed or tempted—everyone agreed about that.

In 1838 (which was way before the Civil War) he refused to sign the Pennsylvania Constitution—because it gave the vote only to white men. People called him an abolitionist—a name that was no help to a politician in that antebellum time. Stevens didn't answer his critics and kept working to end slavery. When fugitive slaves needed help, he was their lawyer, and never charged a fee. He worked hard for free schools, and some wealthy people hated him for that, because it raised their taxes and because they believed ordinary people should be kept ignorant.

If Thaddeus Stevens believed in something, he was willing to fight for it, no matter how unpopular that made him. And what he really believed in were those words of Thomas Jefferson's: *all men are created equal*. His Yankee mind told him that all men meant *all* men —not all white men. So, starting in the 1830s, he began battling for abolition, and then emancipation, and then equal rights. He never stopped fighting, and he never kept quiet.

Neither of the engineers— Johnson (on the left) or Stevens (right)—will give way on his Reconstruction plans.

The Freedmen's Bureau bill of 1866 was intended to make the Bureau stronger and to help former slaves if they were discriminated against. Andrew Johnson vetoes the bill and sends the bureau tumbling, spilling helpless blacks out of its drawers.

Antebellum is Latin and means before (*ante*) the war (*bellum*). In America, it always means the time before the Civil War.

All free governments are managed by the combined wisdom and folly of the people.
—THADDEUS STEVENS

This cartoon depicts Congress attacking Johnson's authority in 1867. It passed a bill, the Tenure of Office Act, that prevented the president from firing his own cabinet members. Johnson's loyal secretary of state, Seward, tries to comfort the stabbed president.

He wrote civil-rights laws. He was the chief author of the 14th Amendment. He laid the foundations for the 15th Amendment (which gave the vote to black males).

Abraham Lincoln was a moderate, a man who believed in compromise. He thought people, at heart, were good. Lincoln had planned a kindly, forgiving Reconstruction. Stevens had no faith in moderation—or in most people. He thought strong laws were needed to make people behave properly.

"We are building a nation," said Stevens, who understood that his ideas would help change a collection of states into a centralized nation.

He believed that the southern states should not be admitted back into the Union until blacks were given the vote, given land, and given guarantees of equality under the law. Those were radical demands.

Thaddeus Stevens was a leader of the Radical Republicans. The radicals became very powerful while Andrew Johnson was president. Johnson hated the radicals. But his hatred helped his political enemies.

President Johnson threw away his popularity. It was that stubborn, narrow-minded streak of his; it made him a poor leader. He didn't listen. He didn't try to represent the whole country. He didn't know how to compromise. He seemed to stand against most Northerners, all blacks, and the moderate southern Unionists. He went on a speaking tour and said wild and nasty things about Congress. Often, he didn't act dignified or presidential. Some people were ashamed of their president.

But Andrew Johnson did have sincere beliefs. He was convinced

How to Impeach

The writers of the Constitution hadn't made it easy to remove a president (or any other high official) from office. This is the process as written in the Constitution (Article I, Section 2):

The House of Representatives shall...have the sole power of impeachment. That means that if a majority of the House of Representatives votes to impeach a president he will be impeached (brought to trial). Then the Constitution says (Article I, Section 3): *The Senate shall have the sole Power to try all impeachments. When sitting for that Purpose, they shall be on Oath or Affirmation. When the President of the United States is tried, the Chief Justice shall preside and no Person shall be convicted without the Concurrence of two thirds of the Members present.*

that it was not the responsibility of the nation to help the newly freed men and women get fair and equal treatment before the law. He thought that was the states' job.

Stevens knew the states had not done that and would not. None of them—north or south.

Thaddeus Stevens detested President Johnson. Their beliefs collided. Johnson called him a traitor and said he should be hanged. Stevens said the president should be impeached.

The Constitution includes a process—called *impeachment*—that allows us to bring to trial federal officials accused of *Treason, Bribery, or other high Crimes and Misdemeanors.* (Look up those words to be sure you know what they mean.) Impeachment is a way to get rid of officials who take advantage of high office and break the law. (It is a process borrowed from the English system of government.)

Thaddeus Stevens said it was "a moral necessity" to impeach President Johnson. Stevens wasn't the only one who wanted the impeachment. Many of this nation's citizens (except those in the South) seemed to want it.

It was 1868. Would the president be impeached? If so, would he then be convicted of high crimes and misdemeanors and thrown out of the White House?

The nation held its breath, waiting to find out.

The cartoonist who drew this Radical Republican express train (Thaddeus Stevens is the driver) thought that it would crash into the hole Andrew Johnson was digging with his veto. And everyone wondered —what was going to happen?

7 Impeaching a President

One of Johnson's "crimes" examined during impeachment was being drunk at the 1865 presidential inauguration.

The debate in the House of Representatives lasted more than two months. Then the House voted. It voted to impeach. After that, it was time for the members of the Senate to try Andrew Johnson. Remember, two-thirds of the Senate needs to vote for conviction in order to remove a president from office. All the senators but one said in advance how they would vote. And two-thirds of the senators—minus one—said they planned to vote against Johnson. The fate of the president would be decided by one vote. That vote belonged to a newly elected senator: quiet, mild-mannered Edmund G. Ross of Kansas.

Ross, a former newspaperman, had been a major in the Union army. He hated slavery and he didn't like President Johnson. He had voted with the Radical Republicans on every issue. Kansas was a

When Stevens was carried to the Capitol for the impeachment, an observer said, "If it were not for the burning fire of his piercing eyes, one would think that life had already forsaken his motionless body."

Radical Republican state, so Senator Ross knew that most of the people he represented couldn't stand President Johnson either. But the young senator said he wasn't going to make up his mind until he heard the evidence.

The day of the trial came, and the chief justice of the Supreme Court administered an oath to each senator, "to do impartial justice."

Picture the scene: a president is on trial. A thousand tickets are printed for admission to the Senate galleries; people do everything possible to get them. One Washington woman wakes a congressman at midnight and won't leave his house until he promises her a ticket. Reporters cover the impeachment as if it were a murder trial. All over the country, day after day, newspaper headlines scream the details.

Edmund Ross gets a telegram from Kansas.

> KANSAS HAS HEARD THE EVIDENCE AND DEMANDS THE CONVICTION OF THE PRESIDENT.

It is signed D. R. ANTHONY AND 1,000 OTHERS.

Ross answers the telegram:

> TO D. R. ANTHONY AND 1,000 OTHERS:...I HAVE TAKEN AN OATH TO DO IM-
> PARTIAL JUSTICE ACCORDING TO THE CONSTITUTION AND
> LAWS, AND TRUST THAT I SHALL HAVE THE COURAGE TO
> VOTE ACCORDING TO THE DICTATES OF MY JUDGMENT
> AND FOR THE HIGHEST GOOD OF THE COUNTRY.

Later, Ross described the scene on the day of the vote.

> *The galleries were packed. Tickets of admission were at an enormous premium. The House had adjourned and all of its members were in the Senate chamber. Every chair on the Senate floor was filled with a Senator, a Cabinet Officer, a member of the President's counsel or a member of the House.*

The vote begins. *Guilty. Not guilty.* Senator James W. Grimes of Iowa, paralyzed from a stroke, has been carried into the Capitol on a stretcher. He votes *not guilty.* Thaddeus Stevens, now an old man, and ailing, is carried in on a chair. He is said to look like "cold marble." He

Sitting second from the left is **Thaddeus Stevens. These were the members of the House of Representatives who managed the impeachment trial in the Senate.**

Ahead of Its Time?

Thaddeus Stevens intended that the 14th Amendment be a weapon against bigotry and injustice. But for almost a century it was ignored or abused. Perhaps if Stevens had been more forgiving, his ideas might have had a better chance in his lifetime. Or maybe he was just ahead of his time. In the second half of the 20th century the kind of civil rights laws Thaddeus Stevens wanted were finally passed and implemented. The 14th Amendment made them possible. Today, the majority of people agree that, aside from the Bill of Rights, the 14th is the most important of all the amendments.

Radical Republican congressman Ben Butler reads out the impeachment evidence against the president.

During the 1864 election campaign (when he was Lincoln's running mate), Andrew Johnson told a group of freedmen, "I will indeed be your Moses, and lead you through the Red Sea of war and bondage to a fairer future of liberty and peace."

Later he told white voters, "This is a country for white men, and by God, as long as I am president, it shall be a government for white men."

A word for that kind of two-facedness is duplicity (doo-PLISS-it-ee).

votes *guilty*. Every senator is present. The chief justice reminds "citizens and strangers in the galleries that absolute silence and perfect order are required."

And then, the moment everyone is waiting for. The chief justice says, "Mr. Senator Ross, how say you? Is the respondent Andrew Johnson guilty or not guilty of a high misdemeanor as charged in this Article?"

Ross remembered it this way:

Not a foot moved, not the rustle of a garment, not a whisper was heard....hope and fear seemed blended in every face. The Senators in their seats leaned over their desks, many with hand to ear....It was a tremendous responsibility....I almost literally looked down into my open grave. Friendships, position, fortune, everything that makes life desirable to an ambitious man were about to be swept away by the breath of my mouth, perhaps forever.

Softly, Edmund Ross says, "Not guilty."

The president is saved.

D. R. Anthony writes to Edmund Ross that *Kansas repudiates you as she does all perjurers and skunks.* Others accuse him of taking bribes. He hasn't.

What he has done is to show tremendous courage. Ross did what he believed was right; he didn't let other people bully him. Later generations will call him a hero, but most people in his time don't understand. Ross is never elected to political office again.

When Thaddeus Stevens hears the vote he says, "The country is going to the devil." Stevens was wrong.

Johnson's mulishness was a disaster for the nation—no question of that. He encouraged racial bigotry. He was a poor leader. He slowed the process of achieving "justice for all." But there were two issues involved, and getting them separated was complicated.

Impeaching a president is a big step. The Constitution says it is to be done for "high crimes and misdemeanors." Johnson was not guilty of that. So it was his ideas that were really on trial. Those ideas were awful—but ideas aren't meant to be impeached or tried. The Founders meant for voters to vote bad ideas out of office.

Most Americans came to believe that the impeachment and trial were mistakes that had threatened the constitutional balance of power. They blamed Stevens and forgot his accomplishments. The 14th and 15th amendments would not have been achieved without him.

Stevens lived for only a few weeks after the trial ended. He spent those weeks writing laws and working on a plan for free schools for the District of Columbia. When he died, his body lay in state in the Capitol. Thousands walked past his casket. Only Abraham Lincoln had received more tribute. He was buried in a cemetery where blacks and whites rest side by side. The words chiseled on his tombstone are his own. They didn't surprise anyone who knew him:

> *I repose in this quiet and secluded spot,*
> *Not from any natural preference for solitude*
> *But, finding other Cemeteries limited as to Race*
> *by Charter Rules,*
> *I have chosen this that I might illustrate in my death*
> *The Principles which I advocated*
> *Through a long life.*

A packed crowd watches the president's trial from the gallery. Andrew Johnson sits scowling crossly in the foreground on the right.

8 Welcome to Meeting Street

Robert Elliott was a black carpetbagger, a Northerner who hoped to change the South.

Eric Foner, *a historian of Reconstruction, writes:* Corruption may be ubiquitous [yoo-BIK-wit-us—it means "everywhere" or "universal"] in American history but it thrived in the Reconstruction South.... "You are mistaken," a Louisiana Democrat wrote a Northern party leader, "if you suppose that all the evils...result from the carpetbaggers and negroes—the democrats are leagued with them when anything is proposed which promises to pay."

Come along to Charleston, South Carolina, and see what is happening. It is January 14, 1868, and we are at the fashionable Charleston Clubhouse, on Meeting Street. The members of the state legislature are about to write a new constitution.

Most of South Carolina's citizens are black. Now that they can vote they have elected black lawmakers to their legislature.

Seventy-six black men and 48 white men are gathered in the ballroom. Outside, beside the palmetto trees, a noisy, colorful crowd surges through the streets. Most are former slaves, and they are full of hope. They believe a new era has begun. Vendors sell benne-seed cookies, shrimp, sassafras beer, and groundnut cakes.

In their handsome, shuttered homes overlooking Charleston harbor, the city's white aristocrats wait and worry. Some of the old plantation families are terrified. Will the assembly demand revenge for the years of slavery? There are rumors that say it will.

At the Clubhouse, a few of the delegates wear rough clothing, but most are dressed in long-tailed frock coats. They wear fashionable whiskers and beards. A Charleston newspaper can't help remarking that "many of the colored delegates [are] intelligent and respectable looking."

Robert Brown Elliott is as well-educated as anyone who is here. He has studied at Eton (a famous school in England). Trained in the law, he is fluent in French and Spanish. Elliott is 25. He will sit in this convention for 14 days without speaking. Then he will stand and quickly become known as a great orator.

At 26, Robert Elliott will go to Washington and enter the House of

Representatives as a congressman from South Carolina. He will work to get civil rights laws passed. This is what he will say in Congress:

> *It is a matter of regret to me that it is necessary at this day that I should rise in the presence of an American Congress to advocate a bill which simply asserts equal rights and equal public privileges for all classes of American citizens. I regret, sir, that the dark hue of my skin may lend a color to the imputation that I am controlled by motives personal to myself....Sir, the motive that impels me is restricted by no such narrow boundary, but is as broad as your Constitution. I advocate it, sir, because it is right.*

When Elliott returns to South Carolina, he will become speaker of South Carolina's assembly. He means to be the nation's first black governor.

But I have gone astray with handsome, clean-shaven Robert Brown Elliott. We need to get back to January 14, 1868. Notice that tall, distinguished man with side whiskers. He is Francis Louis Cardozo, the son of a Jewish economist and a free black woman. Cardozo was graduated from the University of Glasgow, in Scotland, with honors in Latin and Greek. In London he studied to become a Presbyterian minister. He has come to Charleston to be a school principal and a minister. Cardozo is a brilliant administrator. He will save millions of dollars for South Carolina when he exposes corrupt businessmen who are taking advantage of the state. Cardozo is 31.

The man he is talking to also has a Jewish father and a black mother. He is Robert Smalls, and he stunned both North and South a few years ago when, all by himself, he sailed a Confederate steamer through the guns of Charleston harbor and handed it over to the U.S. Navy.

The third man in the group is six-foot-two-inch William Beverly Nash. Nash, a former slave, has taught himself to read. He quotes Shakespeare with ease. Nash is known for his sharp, quick wit.

Franklin J. Moses, Jr., who stands off to the side watching the others, is a likable, well-to-do white Charlestonian. It was Moses who pulled down the U.S. flag at Fort Sumter on the first day of the Civil War. Now some of his old friends call him a traitor and an opportunist because he is working with the freedmen. Moses is charming, but his future is

Robert Smalls lived until 1915 and ended his days as customs collector in Beaufort, North Carolina.

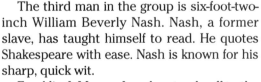

Ever Hopeful

Thaddeus Stevens promised land to the blacks; without it they have almost no power and no chance to better themselves. They must work for others. But, after Stevens dies, the fight for fairness and those inalienable rights gets harder and harder.

Still, the South does seem to be embarking on a great social revolution. In some areas streetcars, schools, parks, and restaurants are integrated. (The South is going further than the North in many situations.)

Despite the early successes of some black legislatures during Reconstruction, for most former slaves the real business at hand was finding a roof over their heads and enough to eat.

Justice for All?

Five years after war's end, black boys and girls attend 4,000 new schools in the South. At least nine black colleges have opened. (They include Fisk and Howard universities and Hampton and Tuskegee institutes.) Black churches are being built in every city and hamlet. Ten years after war's end, Congress passes a civil rights bill prohibiting discrimination in hotels, theaters, and amusement parks. It seeks "equal and exact justice to all, of whatever…race, color or persuasions, religious or political." But, in 1883, the Supreme Court rules that the Civil Rights Law is unconstitutional. (In the 20th century the court will see things differently.)

bleak. He will die a common criminal and drug addict.

There are others to watch here. One is a carpetbagger. He is Daniel Henry Chamberlain, a white man, a graduate of Harvard and Yale, and ambitious. During the war, Chamberlain was an officer of a black regiment. But Robert Brown Elliott will call Chamberlain a "racist" and will hate him. Chamberlain has a keen mind and is an able administrator. He will become governor of South Carolina. Thomas J. Robertson, a former slave owner and one of the richest men in the state, calls the convention to order.

Some say this session is the beginning of "America's Second Revolution." It is an attempt at a genuine interracial society. Poor whites are represented here, and former slaves, and wealthy white men who are accustomed to rule. Some are brilliant; none is stupid. No question about it, this is government by the people. The *New York Times* has sent a reporter. These men are making history—and they know it.

The legislature will soon move to South Carolina's capital, Columbia, and will continue with state business. All across the South, Reconstruction legislatures are at work. Six men in the Florida legislature cannot read or write. (Four of those six are white.) Jonathan C. Gibbs, a black minister and a graduate of Dartmouth and Princeton, is said to be the most cultured man at the Florida convention. Black ministers dominate many of the southern assemblies.

Most Southerners are small farmers. This is a terrible time for them. Cotton prices have fallen to low levels. As if that is not bad enough, the weather is poor and so are the harvests. The white

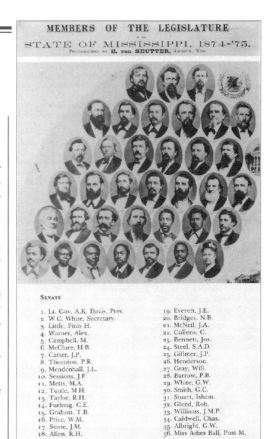

SENATE

1. Lt. Gov. A.K. Davis. Pres.	19. Everett, J.E.
2. W.C. White, Secretary.	20. Bridges, N.B.
3. Little, Finis H.	21. McNeil, J.A.
4. Warner, Alex.	22. Cullens, C.
5. Campbell, M.	23. Bennett, Jos.
6. McClure, H.B.	24. Steel, S.A.D.
7. Carter, J.P.	25. Gillmer, J.P.
8. Thornton, P.R.	26. Henderson.
9. Mendenhall, J.L.	27. Gray, Will.
10. Sessions, J.F.	28. Barrow, P.B.
11. Metts, M.A.	29. White, G.W.
12. Tuttle, M.H.	30. Smith, G.C.
13. Taylor, R.H.	31. Stuart, Isham.
14. Furlong, C.E.	32. Gleed, Rob.
15. Graham, T.B.	33. Williams, J.M.P.
16. Price, W.M.	34. Caldwell, Chas.
17. Stone, J.M.	35. Albright, G.W.
18. Allen, R.H.	36. Miss Adies Ball, Post M.

The members of the Mississippi state senate in its last year of Reconstruction. The photographs are arranged along racial lines. A woman, like Miss Adies Ball, could not be elected to any office, but could be appointed a postmaster.

farmers are exhausted and angry: their sons are dead—killed in the war—their savings are gone. They have no money to hire workers or to buy farm equipment or seeds. The black farmers are mostly farmers without land. It is an impossible situation.

The lawmakers ask themselves: Should they take land from the Confederates who rebelled against their nation? Should they give it to the slaves who have worked the land and made others rich? Can they divide the land and be fair? What about government lands? Can they be given away? How do you give opportunity and justice to all? These are difficult problems—but solvable. Will the Southerners solve them? Will they create a fair interracial society?

These Reconstruction legislatures will vote for free public schools; almost none have existed in the South before. They will vote for roads. They will not demand revenge on the white aristocracy. They will do as well as most legislatures and far better than the United States Congress, which, right now, is shockingly corrupt. But the Reconstruction legislatures will topple. "Redeemer" governments—controlled by former Confederates—will take their place. Fear—for their lives and their jobs—will keep blacks from the polls. The hopeful crowds who swirled about the Charleston Clubhouse in January of 1868 will not see a new era. They will not get to live in a fair interracial society.

Robert B. Elliott addresses the House of Representatives in 1874 about the Civil Rights Bill.

39

9 A Southern Girl's Diary

This was the first racially mixed jury in the southern United States. It was impaneled in 1867 to try Jefferson Davis.

It was seven years after the end of the war, and a Mississippi girl, Mary Virginia Montgomery, wrote this in her diary:

May 1, 1872. A messenger stated that Mr. Jefferson Davis was at Ursino and would come up after Breakfast. We sent the buggy for him. In the meanwhile I brushed around the house, and donned my white dress. Smoothed my hair and pinned on a rose or two. Mr. D. arrived accompanied by Dr. Bowmar. Both gentleman were polished in their manners and on the whole I have been pleased with their visit.

If only we could know what was in Mary Virginia's heart that day. Surely she couldn't have been as calm as her diary sounds. And what did Jefferson Davis think when he met this beautiful, dark-skinned, talented young woman?

Jefferson Davis, you remember, had been president of the Confederacy. When the war was over, in 1865, Davis was a hated man. Southerners needed someone to blame for their defeat. They

Jefferson Davis whines about his prison food to guards who survived the horrors of Confederate prisons.

blamed Davis. Northerners saw him as a traitor. After the war he was imprisoned at Fort Monroe in Virginia. Guards marched in his cell day and night. Then, when he threw his dinner plate at a guard, iron chains were placed on his legs. (The former slave owner complained that he was being treated like a slave.)

Davis had none of the forgiving nature of Robert E. Lee. President Andrew Johnson said he should be hanged. But Congress was full of surprises. The Radical Republicans released Jefferson Davis. Two of the nation's leading abolitionists paid his bail. Thaddeus Stevens volunteered to be his lawyer. Davis wanted nothing to do with Stevens or any Yankee. Besides, his case never came to trial. He went off to England.

Now, seven years after the end of the war, he was in Mississippi, visiting his old plantation. He wanted it back.

It was Mary Virginia Montgomery's home. For the first 14 years of her life she had been the slave of Joseph Davis, Jefferson's older brother. A year after war's end, Joseph Davis sold two of his plantations to Mary Virginia's father, Benjamin Montgomery. He sold them secretly because Mississippi law prohibited blacks from owning land. (A year later the law prohibiting blacks from owning land was overturned and Davis announced publicly that he had sold land to his former slave.)

Joseph Davis had tried to be fair. Before the war he had ruled what he believed was an ideal plantation. He allowed his slaves to run their own affairs, to earn money (after they had done their slave work for him), and to learn and read if they wished.

Benjamin T. Montgomery wished to do all those things. He was a man of unusual ability: a machinist, an inventor, and an able businessman. He managed the Davis plantations and he ran the Davis store. He earned enough money to hire a white man to teach his children. (Until white neighbors complained.) His daughter, Mary Virginia, had a talent for learning. She read books in Latin, studied history, sang, played piano, and rode horseback.

Many people had thought Joseph Davis a model slave owner. But not his slaves. He never gave them what was most important: freedom. They couldn't go where they wanted or do what they wished.

The day after the city of Vicksburg, Mississippi, surrendered to General Grant on July 4, 1863, a photographer took this picture of the former slaves of Davis Bend on their plantation.

Before the Civil War, in 1858, Joseph Davis attempted to patent a boat propeller invented by Benjamin Montgomery. The U.S. Attorney ruled that slaves could not patent inventions because slaves were not citizens. Davis could not patent the propeller either, because slaves could not assign inventions to their owners.

Joseph Davis

In 1887, Isaiah Montgomery founded a black town he named Mound Bayou. In 1895, Mary Virginia became postmistress of Mound Bayou.

Mary Virginia Montgomery's parents, Benjamin Thornton Montgomery and Mary Lewis Montgomery. Mrs. Montgomery was in charge of cotton production at Davis Bend.

The black community of Davis Bend outside Hurricane Garden Cottage—the humble name for Jefferson Davis's plantation home.

So when the Union army came into Mississippi and Joseph Davis fled, his slaves didn't go with him. They went off in different directions. But Benjamin Montgomery soon came back to the old plantations at Davis Bend. General Ulysses S. Grant helped him rent the Davis plantations. Montgomery advertised in a black newspaper (those papers were springing up all over the South). He wrote that he intended to "organize a community composed exclusively of colored people." He said he was looking for people of "honesty, industry, sobriety and intelligence." Freedmen and freedwomen flocked to Davis Bend. In 1865, laborers there raised 2,000 bales of cotton and earned a profit of $160,000.

Everyone in the Montgomery family worked to pay for that land. Mary Virginia was a bookkeeper and helped in the post office. Brother William was a U.S. postmaster. Brother Isaiah managed the store. Sister Rebecca sewed and worked in the store. Their mother was in charge of cotton production. In 1870, Montgomery cotton won a prize as the best at the St. Louis Fair. Six years later, it was best at the national centennial.

Davis Bend was now a community of 1,600 people (only 30 of whom were white). It had its own judges and sheriffs and political leaders.

Some in the community wanted Jefferson Davis released from prison. "Although he tried hard to keep us all slaves…some of us well know of many kindnesses he showed his slaves on his plantation."

Meanwhile, Mary Virginia and Rebecca Montgomery went off to Oberlin College. Just like many other college girls, they were homesick at first. This is part of what Mary Virginia wrote about college life:

> *December 3. This day, my most fearful anxieties have been calmed. The examination is passed. It was not so rigid as I expected. Prof. Smith delivered…kindly words of welcome [that] sank deeply in my poor homesick heart.*

Mary Virginia Montgomery studied hard and did well at college. Then she and her sister came home and became schoolteachers in Davis Bend. By this time Benjamin Montgomery had bought another plantation. It was

Jefferson Davis's former home, Ursino. The Montgomerys were now said to be the third largest planting family in Mississippi. But times were changing. Benjamin Montgomery died. So did Joseph Davis.

Jefferson Davis survived. He was different from his older brother. He had never had to work to earn money. It was his brother's earnings that had made him rich. He was used to having them. He was used to having people work for him. He wanted the plantation back. He wanted his old home. He soon had power on his side. Reconstruction failed; laws were changed; most blacks lost the freedom to vote; and, once again, Jefferson Davis became the hero of the white South. He got what he wanted, and Mary Virginia had to move.

A black candidate for public office stumping in the South.

Life, Liberty, and Property Unprotected

The following petition was made to the U.S. Congress on March 25, 1871:

We the colored citizens of Frankfort and vicinity do this day memorialize…upon the condition of affairs now existing in this state of Kentucky. We would respectfully state that life, liberty, and property are unpro-tected among the colored race of this state. Organized bands of desperate and lawless men, main-ly composed of soldiers of the late Rebel armies, armed, disci-plined, and disguised, and bound by oath and secret obligations, have by force, terror, and violence subverted all civil society among the colored people….We believe you are not familiar with…the Ku Klux Klan's riding nightly over the country, going from county to county, and in the county towns spreading terror wherever they go by robbing, whipping, ravishing [raping], and killing our people without provocation, compelling colored people to break the ice and bathe in the chilly waters of the Kentucky River….Our people are driven from their homes in great numbers….We would state that we have been law-abiding cit-izens, pay our tax, and, in many parts of the state, our people have been driven from the polls—re-fused the right to vote. Many have been slaughtered while attempt-ing to vote; we ask how long is this state of things to last. We ap-peal to you…to enact some laws that will protect us and that will enable us to exercise the rights of citizens….the senator from this state denies there being orga-nized bands of desperadoes in this state…we lay before you a number of violent acts occurring during his administration.

White Democratic thugs frequently intimidated black Republican voters.

43

10 A Failed Revolution

An Alabama paper ran this cartoon, both racist and anti-Republican, titled *A Sample Grant Voter.*

Successful revolutions are rare.

During the Reconstruction period black lawmakers voted for schools, roads, and railroads. Schools, roads, and railroads are especially helpful to the average person. However, you need to levy taxes to pay for them, and taxes cost big landowners more than they cost the average person. Do you see a problem here? The landowners were not happy at all about some of the new laws.

Besides, there was another problem at the bottom of all this. If blacks could be congressmen and responsible citizens, then the whole idea of black inferiority didn't make sense. So slavery must really have been wrong. Many white Southerners still couldn't accept that idea. They had been through a ghastly war. They had lost their loved ones. If slavery was wrong, their sons and fathers had died for nothing. How could they believe that? White Southerners were not monsters; they were not angels; they were human beings.

In the South, a new form of farming developed after the war. It was called sharecropping. A landowner supplied land, tools, and seed to a landless farmer, who then gave the owner one third or one half of all he grew. At first it seemed a fair system, but it rarely turned out that way. After sharecroppers had paid the landowner they usually had almost nothing left for themselves.

President Grant loved to drive. He was once stopped by a Washington policeman for going too fast.

Grant struggling with the country's problems. A newspaper editor wrote, "It is a political position and he knows nothing of politics."

They needed some help understanding what had happened to them and to their beloved South.

Abraham Lincoln, who might have led them wisely, was gone, killed by an assassin's bullet. The former Confederates hated the Radical Republicans—and, as you know, Andrew Johnson was the wrong man as president. So was the next president. This may come as a surprise to you, but Ulysses S. Grant—General Grant, the hero of the Union, the powerful warrior, who had once been compared to the Greek hero, Ulysses—was now sometimes called Useless S. Grant. Being a great general had made him a popular hero and gotten him elected president, but it had not prepared him for political life.

Most Americans loved President Grant. And there were good reasons for that. He was a fine man: honest and honorable. But he was too trusting. Grant trusted the men around him; unfortunately, they turned out to be untrustworthy. They stole from the nation. They took millions and millions of dollars in public lands and resources. Ulysses Grant, and the American people, were their victims. The Grant years were years of terrible corruption.

All that corruption made many white Americans forget about civil rights for others. By the time Grant left the presidency, in 1877, the North's citizens were tired of hearing about the need for a just society in the South. They had their own problems, and when they thought about it, they realized their society wasn't perfect either. Maybe they should leave the South alone. When hate groups began blaming immigrants and blacks for the nation's problems, many listened. Then, when Rutherford B. Hayes promised to pull the federal troops out of the South if he became president, he got that job. And he kept his promise.

The old guard in the South began to take power again. And they didn't worry about sharing it. They didn't worry about justice for all. They passed laws that made voters pay a *poll tax*: that meant most blacks couldn't vote. They made it impossible for blacks to get a decent education or to buy land. They would not allow blacks to have fair trials. Soon many Southern blacks were not much better off than they had been when they were slaves. Some were worse off. It was clear: Reconstruction was failing.

All of that still wasn't enough for some white racists. So they

Among other things, 1877 marked a decisive retreat from the idea, born during the Civil War, of a powerful nation state protecting the fundamental rights of American citizens.

—ERIC FONER

What does Foner mean? Was the idea reborn? When? (Keep reading U.S. history and you'll find out.)

Poll is an old word for "head." So a poll tax is a tax you pay just because you're a person. Most blacks couldn't afford to pay the tax, and if you didn't pay the tax you weren't allowed to vote.

45

Lynching means to execute without due process of the law, especially by hanging. In simpler words: lynching is murder, usually murder by a mob. (More on lynching in Chapter 34). This pro–Ku Klux Klan cartoon in an Alabama paper sketched the likely fate of a carpetbagger if the Democrats won the 1868 election.

In Virginia, a few blacks continued to be elected to state office until 1901, when Virginia held a convention to write another state constitution. The new constitution took the vote from most blacks and most poor whites. Now Reconstruction was really dead and done with. It had failed in its goal: to make the South into a just society where people of all races could live in harmony.

You can read about Jim Crow in Chapter 32 of this book.

joined terrorist organizations (like the Ku Klux Klan) and lynched and intimidated blacks who wanted to vote.

The good people, who wouldn't do those things, didn't shout out. They didn't stop the bad actors. Most of the Northerners who had tried to help in the South got discouraged and left. Some of them were lynched. The Southern whites who tried to be fair were ridiculed—or lynched. Slowly, a policy of separation of races was instituted—it was called *segregation*. Blacks were not allowed to sit on the same seats that whites used. They were not allowed in the same restaurants. They had to sit in the back of public streetcars. They could not go to school with whites. State and local laws were passed that made all those unfair things possible. No one did anything about them. The Supreme Court didn't seem to care—sometimes it sided with the racists. Whites and blacks who tried to fight for equal rights were forlorn. There was no leader big enough to reach people's hearts.

When President Rutherford B. Hayes did what he had promised—he pulled the soldiers out of the South—congressional Reconstruction was just about finished. In Arkansas, men with guns made black lawmakers leave the state. In most southern states blacks were prevented from voting. Robert Brown Elliott now had no hope of being South Carolina's governor.

In his book *The Promise of the New South*, the historian Edward L. Ayers writes:

Rutherford Hayes kissing a baby on his presidential campaign. He won through fraud: the votes in Louisiana and Florida were fixed.

> *Conservative governments opposed to Reconstruction took over in Virginia in 1869, in North Carolina in 1870, and in Georgia in 1871. Democrats regained dominance in Texas in 1873, in Alabama and Arkansas in 1874, and in South Carolina and Mississippi in 1876. Reconstruction's final gasp came in 1877, when Congress declared victory for the Democrats in contested elections in Louisiana and Florida.*

Reconstruction wasn't totally over in 1877, but it was close to it. To make things worse, a Northerner, a fool named Jim Crow (a clown character who believed in separating the races), had come south.

That policy of white supremacy didn't help either the black or the

Governing South Carolina

It was at the battle of Gettysburg; a Michigan soldier and a South Carolina cavalryman were shooting at each other when the Northerner's gun jammed. Wade Hampton, the Confederate, waited courteously until his adversary had fixed his rifle. Then he shot him through the wrist.

No one who knew Hampton was surprised by that story. His courage was legendary; so was his great physical strength—and his courtesy. Before the war, he may have been the richest man in the South; his family's plantations—in three states—were worked by 3,000 slaves. After the war, he had almost nothing left, but he didn't seem bitter. "I have claimed nothing from South Carolina," he said, "but a grave in yonder churchyard." He supported the idea of votes for blacks—the first leading Confederate to do so.

Wade Hampton

And he attempted to be fair and honorable to all people—but his vision of fairness included an idea of fatherly responsibility toward the former slaves. Wade Hampton couldn't get it into his head that blacks could be his equal. Maybe he had lived with slavery for too long.

Hampton hated the Reconstuction governments; he didn't want those who had opposed his beloved Confederacy telling him what to do, so, in 1876, when the Democrats urged him to run for governor, he accepted. But when groups of thugs, called "Redshirts," decided to help the election along using their guns, Hampton spoke out against their methods, although he agreed with them about white supremacy. After the voting, both sides claimed victory. Federal troops, and President Grant, supported Republican

governor D. H. Chamberlain (Robert Elliott was elected attorney general, and F. L. Cardozo, treasurer). For three months, South Carolina had two governors and two administrations. Then President Hayes took over. Hayes was an honest man, although voter mischief in some southern states put him in office. He agreed to pull federal troops out of the South after being promised that blacks would be treated fairly. The promise wasn't kept. Without troops, Governor Chamberlain could not control the Redshirts. Wade Hampton became governor. Reconstruction was finished in South Carolina.

Governor Hampton was as fair and gentlemanly as he knew how to be. While he was in office—as governor and then as U.S. senator—blacks had the vote in South Carolina. But in the 1890s, when a bigot named Benjamin Tillman took over as governor, the race haters were in control.

white South. It did just the opposite. Industry and new ideas abounded in the North and West. But until fairness began to return (after World War II) immigrants and most big industries stayed away from much of the South. Southern industry consisted of foundries, carpentry shops, and small manufacturing plants. Tobacco and cotton were still the major crops. Wages were low. The South became the poorest section of the nation. Southern students ranked at the bottom of the national charts. (This is where studying history helps you understand some things. Poor leadership cost the South prosperity. It limited opportunities for all its citizens.)

But don't be too hard on the South. A fair-minded interracial society didn't exist anywhere in the world in the 19th century. Not in the North,

Buses in the 19th century? Yes, they existed—they were often double-deckers—but of course they were pulled by horses.

No Redeeming Features

The Republican Party was the party of Reconstruction. The Democratic Party (in the South) was determined to bring back as much of the old South as possible. Conservative Democrats called themselves *redeemers*. The Redeemer Democrats had no interest in working with blacks. They were not concerned with black needs, and they certainly didn't want blacks to hold office. Nor did the Democrats approve of the idea of an active government trying to improve conditions for all people. The Republicans wanted the state governments to support railroads and businesses and to improve schools and welfare institutions (orphanages, prisons, etc.). The Redeemer Democrats laughed at those ideas (they all cost money). The Democrats wanted to lower taxes. They redeemed one state after another, driving the Republicans from power. (Remember, this was the 19th century; the parties are very different today.)

not in the West, not in Europe, not in Asia, not in Africa, not anywhere. In the United States a first step had been taken. That first step had been in the South during Reconstruction. And that time had also produced those three remarkable amendments: the 13th, 14th, and 15th.

In the second half of the 20th century a black minister would help heal the old wounds and begin real reconstruction in America. And then, finally, America would have its first black governor. He would be elected in Virginia.

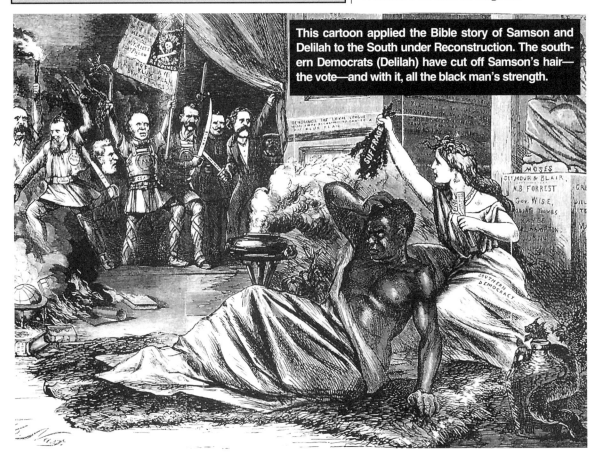

This cartoon applied the Bible story of Samson and Delilah to the South under Reconstruction. The southern Democrats (Delilah) have cut off Samson's hair—the vote—and with it, all the black man's strength.

11 Meanwhile, Out West

GRAND RUSH
FOR THE
INDIAN
TERRITORY!

Over 15,000,000 Acres : Land
NOW OPEN FOR SETTLEMENT!
Being part of the Land bought by the Government in 1866 from the Indians for the Freedmen.

NOW IS THE CHANCE
TO
PROCURE A HOME

This poster advertises "part of the land bought by the government in 1866 from the Indians for the freedmen."

There was war in the West. It was the new settlers fighting the Indians, the buffalo, and nature for control of the land. Buffalo are dumb, but even if they'd been smart they didn't have a chance. And nature—well, nature is tough, but no match for people. Six-hundred-year-old trees got axed. Big areas—like the Wisconsin Cutover—were left nude and unable to grow anything. A few people were beginning to worry about the environment, but hardly anyone listened to them.

As for the Indians, that was a different story: they were intelligent and resourceful, but they were outnumbered and outgunned. They didn't have much of a chance, either. The settlers had been fighting wars with the Indians since the Pequot War, back in Massachusetts in 1631. The wars—no matter what people said—were all over the same thing: land and who would control it. The Native Americans had been promised—in treaty after treaty—that if they would just move once more, they would be left alone. First they were asked to move across the Appalachians; then west of the Mississippi; now it was off all the good land left in the West. Well, they wouldn't do it. But they weren't given a choice. At least, it wasn't much of a choice: it was move or fight. And that led to the 30 years of Indian wars in the West. You'll hear more about them—the outcome wasn't good for the Native Americans.

It is hard to blame the people who took their land. They were just ordinary people who had been told there was free land in the West. So they'd

In 1869, geologist John Wesley Powell, who lost an arm during the Civil War, heads an expedition that goes down the Colorado River and through the Grand Canyon. Do you want to read an adventure story? Find out about that trip.

In Guthrie, Oklahoma Territory, two swaggering land-rush settlers have put up a tent and are guarding their claim.

A photographer was on hand to shoot history's greatest land rush: in 1893, 6 million acres of "unoccupied" Oklahoma land were claimed in a day.

A homesteader turns over virgin prairie in Montana.

come to settle on it. Many had also been told awful, untrue stories about the Native Americans. They thought all Indians were savages and cruel and monsterlike. They were terrified of Indians and wanted them all to disappear. And some Indians were killing innocent people.

Besides, the settlers weren't having an easy time. Part of the problem was an economic depression in the East. It came on suddenly, and thousands of people were out of work. Many of them headed West, across the Mississippi.

But they found farming in the prairie lands different and difficult—much harder than farming back East. Plains farmers needed to be able to stand up to extremes of heat and cold, and to violent storms. There was something else about this kind of farming that was important—and new to the Easterners: a big investment of capital (which means money) was needed for farm equipment and seeds. Many farmers weren't prepared. IN GOD WE TRUSTED, IN KANSAS WE BUSTED, said signs on the covered wagons heading back east. About one third of the people who came west turned around and went back home.

But, despite the hardships and the roller-coaster economic conditions, mostly there was optimism. The midsection of the nation began to produce so much wheat it was called "the nation's bread basket." Western ranchers were raising cattle and sheep to put steaks and chops on people's dinner plates. Western miners were digging up important minerals. Railroads could take those products to far places.

In the 1870s, no one knew quite where the nation was heading, but it was easy to see it was going in new directions—and fast.

Cities were springing up everywhere: Ohio's cities were among the fastest growing. Chicago was the nation's busiest port. (Chicago a port? In the middle of the country? Look at a map and see how that could be.) Factories spurred the growth in most cities; they produced affordable goods, and jobs, too—especially for the new immigrants. America's citizens were beginning to have comforts and possessions beyond anyone's dreams.

Presidents and congresses couldn't seem to keep up with all the new ideas, opportunities, and problems. But who could?

Before the Civil War the United States had been a midget in world affairs. Now it was in a growth spurt. Before long the nation would be an industrial and agricultural giant.

The poet Walt Whitman described what was happening:

> Land of coal and iron! land of wool and hemp! land of the apple and the grape!
> See, in my poems, cities, solid, vast, inland, with paved streets, with iron and stone edifices, ceaseless vehicles, and commerce.
> See, the many-cylinder'd steam printing press — see, the electric telegraph stretching across the continent....
> See, the strong and quick locomotive as it departs, panting, blowing the steam whistle,
> See, the ploughmen ploughing farms — see, miners digging mines — see, the numberless factories,
> See, mechanics busy at their benches with tools — see from among them superior judges, philosophs, Presidents, emerge, drest in working dresses.

When there weren't any trees, you burned dry buffalo chips. A Kansas woman brings home a barrowful. What are buffalo chips?

Nebraska became a state on March 1, 1867; Colorado on August 1, 1876.

Lincoln, Nebraska, in 1872, five years after statehood. The capital city was a tiny island in a vast sea of prairie.

12 Riding the Trail

Range is open land for grazing cattle.

"O say, little dogies, when you goin' to lay down, And quit this forever shifting around?"

Some say Elijah McCoy was the "real McCoy." Elijah was the son of a runaway slave. He went to Scotland and became a mechanical engineer. Then he came back to the United States, where he invented a lubricating cup that fed oil into machines as they operated. He invented many other things too. He held 75 patents.

Another "real McCoy"? Bill McCoy was a rum runner during Prohibition. His product was the real stuff. (If you want to know about Prohibition, read Book 9 of *A History of US*.)

After the Civil War, when soldiers came home to Texas, they found the place swarming with longhorn cattle. Now longhorn, as the name tells you, are cattle with long horns growing out of their heads. They are a tough breed and can walk great distances. And, if there is grass to chew, they can even fatten up on the journey.

The Texas longhorn were descended from cattle brought to America by Columbus and the Spaniards who followed him. The longhorns were running loose on the range. There they bred and multiplied and were soon so numerous that people were killing them for their hides and throwing away the meat. The ex-soldiers knew beef was expensive back East. Now, if they could find a way to get those cattle east—why, there was money to be made.

About this time, Jesse Chisholm—who was half Scot and half Cherokee—drove a herd of cattle north from Texas to Kansas and made a map of his route. That route had plenty of grass for grazing and enough water and it led to Abilene, Kansas. In Abilene, Joseph G. McCoy was paying $40 a head for cattle. Forty dollars was a powerful lot of money in those days, especially since you could

Jesse Chisholm

get longhorn in Texas for about $5. If you had a herd of 1,000 or more steers and cows and got them to Abilene—well, you can figure out that you'd be rich.

The Kansas Pacific Railroad reached Abilene in 1867, so Joe McCoy could ship the longhorns east in

Joe McCoy

railroad cattle cars and make a lot of money for himself. Which he did. He made a whole lot more money than even he expected. And he became famous. Have you ever heard of the real McCoy? Some say that was Joe.

Remember Chicago? The little town where Abe Lincoln campaigned? Well, most of Joe McCoy's cows and steers got shipped to Chicago. That city—some people call it the Windy City—was both a port and a railroad center. It was a distribution hub. When refrigerated railroad cars were developed, Chicago became the meatpacking capital of the country. Most of the bovines that traveled Jesse Chisholm's trail got turned into steaks in Chicago. Thanks to railroads and ships, they were devoured in Norfolk and Scranton and Mobile.

For the next 20 years the Chisholm Trail (that's what they named it) was a ribbon of longhorn. More than a million cattle were driven north on the trail. Who drove the herds? Why, cowboys, of course. What did the cowboys do when they got to Abilene? Get paid, buy fancy new duds, and have a big old time. Did they get rich? Not the cowboys; they were paid about $90—or less—for the two-to-three-month journey from the Texas panhandle, across the Red River, through Indian territory, over deserts, rivers, and prairie. It was the cattlemen, who owned the cows and steers, who got rich, but that's another story.

What the cowboys got was a way of life that turned them into legends. Some people call cowboys "knights of the prairie." And they were like knights: they rode with amazing skill, handled danger with bravado, and had their own code of honor. It may sound glamorous, but it was not an easy life. There were killers on the Chisholm Trail; here are some of them:

brutal heat,
ferocious blizzards,
biting hail,
angry Indians,
rattlesnakes,
quicksand,
rustlers,
thirst—and,
most common of all, **stampedes.**

A herd of cattle will stampede at the drop of a frying pan. Here is how one cowboy described it: "While I was looking at

Nat Love, a black man from Tennessee, became famous as Deadwood Dick, one of the great cowboys. He took part in many cattle drives.

Western Wisecracks

Q. Why did the cowgirl eat a box of bullets?

A. She wanted her hair to grow bangs.

Q. What did the blanket tell the U.S. marshal?

A. Don't move, I've got you covered.

Q. What is a sick longhorn?

A. A bum steer.

Show Time

Cowboys liked to be entertained, so stage shows often made their way west. But the entertainers had to be prepared for gun-slingers who sometimes got excited and pulled out their weapons. At a performance of *Uncle Tom's Cabin* in the Bird Cage Theater in Tombstone, Arizona, one cowboy was so upset at seeing the blood-hound chase poor Eliza that he fired his gun at the dog, killing both the animal and the show. The curtain went down, and the disappoint-ed audience had to be restrained from stringing up the gunman.

Comedian Eddie Foy was performing in Dodge City when some cowhands lassoed him and took him to the local "hanging tree." They were angry about some jokes he'd made about them onstage. Foy asked if he could say his last words at a well-known saloon—and that act of audacity won the cowboys over. After that they were his best audience.   *The audience gets carried away.*

him, this steer leaped into the air, hit the ground with a heavy thud, and gave a grunt that sounded like that of a hog. That was the signal. The whole herd was up and going—and heading right for me. My horse gave a lunge, jerked loose from me, and was away. I barely had time to climb into an oak. The cattle went by like a hurricane, hitting the tree with their horns. It took us all night to round them up."

That cowboy was lucky there was a tree to climb and the cattle didn't knock the tree over. If they had, it would have been goodbye cowboy.

Cowboys were usually up before dawn and were often still hard at work into the night. But there was something about the life that most of them loved. As one cowhand said, "To ride around the big steers at night, all lying down full as a tick, chewing their cuds and blowing,

"An endless grind of worry and anxiety that only a strong physical frame could stand," said one cowboy of the trail.

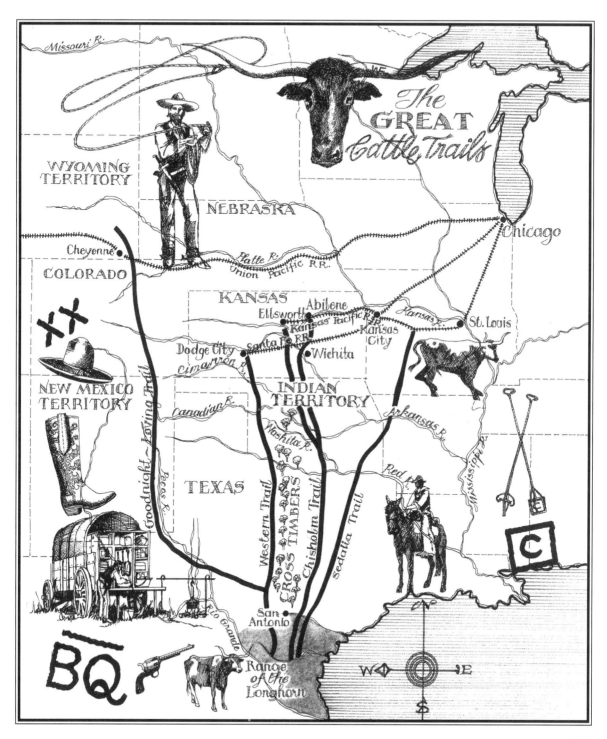

The GREAT Cattle Trails

In the Kansas cattle towns there were twice as many saloons as other businesses. On the trail, grub was dished up from the chuck wagon. It had shelves and drawers for coffee and flour, beans and sugar. A swinging leg held up the open box lid and turned it into a tabletop.

When the cowpokes talked about the weather —well, they had some stories to tell. Did you know that when the wind blows over the prairie and a chicken turns its tail into the wind, it sometimes lays the same egg three or four times?

In really cold weather, cowboys say their words freeze and, weeks later, travelers sometimes hear an outburst of words that seem to come right out of the air.

with the moon shining down on their big horns, was a sight to make a man's eyes pop."

Pretty soon it got to be a regular thing—traveling the Chisholm Trail. Herds of two or three thousand cattle became common. Usually a dozen cowboys were hired to handle a herd, with a trail boss and cook. The cook was important. Cowboys got ornery if the coffee wasn't strong and the food decent.

Like other Americans, cowboys were a mixture: some white, some black, some Mexican, some—like Jesse Chisholm—part Indian. Some were women. It was democratic out there on the trail. People were judged by what they could do, not by the color of their skin, the accent in their speech, or their sex.

Elizabeth E. Johnson was a schoolteacher in Texas before she bought some cattle and became a cowgirl. In 1879, she drove her own herd up the Chisholm Trail to market. When she died, 45 years later, she was worth more than $2 million.

Most cowhands wore tightfitting clothes, leather chaps, floppy vests, fancy boots, and broad-brimmed hats—clothes adapted from those of the cattlemen who had come to the western land from Spain. Much cowboy lingo was Spanish: *chaps, lariat, rodeo, ranch* (and *lingo*, too).

It was a lonely life they led, so they livened it up by singing around the campfire, or telling tall tales. Like the one about the cowboys who ran out of fenceposts during a snowstorm when they needed to fence the herd in.

In 1876 Wild Bill Hickok (left) was killed in a saloon in Dakota Territory. Center: Bat Masterson, who died a successful New York journalist. Right: Wyatt Earp, sheriff, who moved to Los Angeles and became a friend of movie-star cowboy Tom Mix.

So they hammered frozen rattlesnakes into the ground and strung wire on them. Which would have worked, except there was a thaw, and the fence just crawled away.

And Abilene—what happened to Abilene? It was the first of the wild-west towns—and maybe the wildest of them all—with saloons and pistol-packing cowpunchers raring for a good time. A town like that needed a marshal, and Abilene got the most famous one of all: James Butler ("Wild Bill") Hickok. Hickok had once been a bad hombre himself, a gambling man who had two pearl-handled pistols and was known as the "fastest draw in the West." He could shoot the hat off a man and keep the hat in the air with his bullets and when it finally dropped it would be rimmed with a circle of bullet holes. At least that was the story he told. Hickok was a sharp dresser who wore a coat with satin lapels, parted his long hair in the middle, and slicked it and his mustache with bear grease.

Wild Bill was paid a big salary—$150 a month—for keeping law and order in Abilene. He did a fair job for a while, although most of the time his office was at a gambling table. Then he shot two people; he'd done that before, but it just happened that one was his deputy. That was too much for the good folk of Abilene. They got rid of Hickok and the cattle market, too. Abilene settled down and became respectable. After that the cowboys and their longhorns headed for the new towns of Wichita and Dodge City. About 1875, Dodge City became the main railhead for shipping cattle eastward. Most say it was an even wilder place than Abilene (and its "peace" officers—Bat Masterson and Wyatt Earp—well, you should read about them). But by the end of the century, when railroads crisscrossed the land, and barbed wire fenced it in, the heyday of the wild cowtowns was over.

More Than Just a Plain Jane

No one ever knew how many of Calamity Jane's stories were true and how many were tall tales. No one much cared. She was a good storyteller and no ordinary woman—even in those frontier days.

She was Martha Jane Cannary and, at age 13, already an expert rider when her family headed west. It was a five-month trip, and she spent it exploring and hunting. It was a good thing she did. When her parents died—a year and a half later—she could take care of herself. Calamity Jane claimed to have been a gold miner, a nurse, a Pony Express rider, an army scout, an Indian fighter, and a cattle hand.

Now that is what she said. Some people say all of it is true and some say none of it is—including her name. No question, she was a well-known woman of the Wild West, and a friend of Wild Bill Hickok—and a drunk. How did she get her name, Calamity Jane? Well, there are so many stories about that, it isn't worth the bother to tell them. (Do you know what calamity means? It describes her.) We do know that at 23 she joined a geological expedition to the Black Hills, and at 24 she was the only woman, among 1,500 men, in an expedition that headed out from Fort Laramie to fight the Sioux.

13 Rails Across the Country

This all-time record was the result of a bet between two of the railroad owners.

It was May 10, 1869, and hardly a person in the whole country didn't wish to be at Promontory Point in Utah. The people who were there listened to speeches, said a prayer, drank toasts (too many, so it was said), yelled, and cheered. Two brass bands blared. All over the country, newspapers held their presses so they could cover the grand event. The newspapers wouldn't tell the whole story—it would take years for the real story to be known—but when Leland Stanford (representing a railroad company that had laid tracks east from Sacramento, California, over the high and dangerous Sierra Nevada mountains and on to northern Utah)—when Leland Stanford shook hands with Thomas Durant (whose railroad company had laid tracks west from Omaha, Nebraska, over land the Indians thought they owned)—why, the whole country got excited. A telegraph operator, on a high pole above the crowd, sent out the message STAND BY, WE HAVE DONE PRAYING. Then Leland Stanford raised a silver hammer, whomped at a solid gold spike—and missed. No matter, the next swipe hit the nail on its head.

Five states had sent gold and silver spikes for this historic event, so the other bigwigs each got a chance to hammer away. When they were done, Chinese workmen quietly removed the fancy spikes and nailed regular ones in place. By that time the news had been sent to America's newspapers and people. TRANSCONTINENTAL RAILROAD COMPLETED. EAST AND WEST LINKED. In Philadelphia the Liberty Bell rang. Chicago held a parade that stretched for seven miles. In New York cannons blasted 100 times. You could now go by train from New York to California.

The building of that railroad had begun, in a way, on July 1, 1862, the day Abraham Lincoln signed the Pacific Railroad Act. There were visionaries—people with clear sight and imagination—who had talked even earlier of a railroad to cross the country. Lincoln's act got it started. But slowly. After all, there was a Civil War being fought. The country learned of the importance of railroads during that war, when armies were moved by train. But a railroad that stretched across the continent—that would have to wait until the war was over.

Two companies built the railroad. The Central Pacific (starting in the West) and the Union Pacific (coming from the East). No one knew exactly where they would meet. It became a race—East against West. It was an important race for those who owned the railroad companies. The government was giving subsidies for each mile of railway track that was laid. So, of course, each side wanted to lay the most track.

Under the best conditions, laying track isn't easy. Conditions were rarely "best." Remember, this was frontier land they were crossing. The railroads had to bring all their supplies with them. If there was an emergency there was no place to go for help.

You lay tracks by putting heavy metal rails on top of wooden cross pieces—trimmed logs—called "ties." The Union Pacific used 40 railroad cars to haul the 400 tons of rails, timber, fuel, and food needed for each mile of track. The Central Pacific brought its rails, locomotives, and supplies from the East Coast on clipper ships that sailed around to Cape Horn at the tip of South America.

Laying tracks on flat land is not too difficult. But try crossing a mountain. You have a choice: you can go over or through. The railroad men did both of those things—they dug tunnels through some mountains and laid tracks over others. At first the equipment they used was about as fancy as what you might find in a neighbor's garage. Workers attacked rocks with pickaxes, they dug tunnels with shovels and their bare hands. They carried stones and dirt in wheelbarrows. Sometimes they used explosives; sometimes they blew themselves up.

After the golden spike was driven home, the Union Pacific engineer, General Grenville Dodge (left), a distinguished Union veteran of the Civil War, shook hands with the Central Pacific engineer, Samuel Montague.

A **subsidy** is a grant of money, land, or something of value. The railroads got valuable land grants.

59

The Central Pacific crews competed with the Union Pacific crews to lay the most track—until the government chose Promontory as the specific meeting place and settled the question.

Do you want to build a railroad, 19th-century style? Start with *graders*: men with picks, shovels, wheelbarrows, and wagons—they grade the land, making it as level as possible. Muscular *tracklayers* follow. They lift and place the heavy wooden ties and the heavier metal rails. Next come *gaugers, spikers,* and *bolters*—who get it all together and hammer the spikes in place. Each rail takes 10 spikes and each mile 400 rails.

The Central Pacific Railroad sent ships to China and brought 7,000 Chinese workers to California just to build the railroad. They paid them $1 a day. The Chinese worked incredibly hard, for long hours, and, mostly, were treated with contempt.

Coming the other way, on the Union Pacific, most workers were either ex-Confederate soldiers, former slaves, or Irish immigrants. They, too, worked hard. They lived in tent cities put up and taken down as the railroad went west. Those moving towns were tough, violent places where there was too much drinking and too many guns. There was constant fear of Indian raids.

The Indians—by treaty—had been guaranteed land west of the 95th meridian (a merid-

A Thousand Yippees

In a book called China Men, *Maxine Hong Kingston writes about her ancestors, who came from China and helped build a railroad:*

There were two days that Ah Goong did cheer and throw his hat in the air, jumping up and down and screaming Yippee like a cowboy. One: the day his team broke through the tunnel at last....The second day the China Men cheered was when the engine from the West and the one from the East rolled toward one another and touched. The transcontinental railroad was finished. They Yippee'd like madmen. The white demon officials gave speeches. "The Greatest Feat of the Nineteenth Century," they said. "The Greatest Feat in the History of Mankind," they said. "Only Americans could have done it," they said, which is true. Even if Ah Goong had not spent half his gold on Citizenship Papers, he was an American for having built the railroad. A white demon in top hat tap-tapped on the gold spike and pulled it back out. Then one China Man held the real spike, the steel one, and another hammered it in.

While the demons posed for photographs, the China Men dispersed. It was dangerous to stay. The Driving Out had begun.

It is the winter of 1868, and Union Pacific crews have just crossed Wyoming's Green River in the shadow of Citadel Rock and the Wasatch range. On the other side of the mountains lie Utah, the Great Salt Lake, and Promontory Point.

ian is a line of longitude) as a permanent home. When the railroad got to Nebraska it was already at the 100th meridian. So Indians, understandably, were angered. They hated the thundering locomotives that were destroying the buffalo ranges. Native Americans raided the railroad camps, but not as often as the tales say. Indians didn't kill many trainmen; it was disease, accidents, avalanches, heat, and cold that were the worst killers. Some people deserved medals for what they accomplished—expecially the engineers and organizers who arranged for the food and shelter, who planned the route, and who supervised the work. None of that was easy. They did it with incredible speed.

After the railroads met at Promontory Point, everyone had to wait two days to celebrate. That was because Thomas Durant was late. Union Pacific workmen had chained his fancy parlor-car train to some railroad track. They kept him hostage until he paid them overdue wages.

By that time Durant was rich enough to make King Midas envious. So was Leland Stanford, who made his employees salute when he rode

A couple were married, while the cars were going at the rate of 30 miles per hour. Marital matters and things are becoming of such frequent occurrence on our railways that we suggest the keeping of train ministers and doctors to meet cases of emergency.

—*NEBRASKA HERALD,*
JANUARY 12, 1871

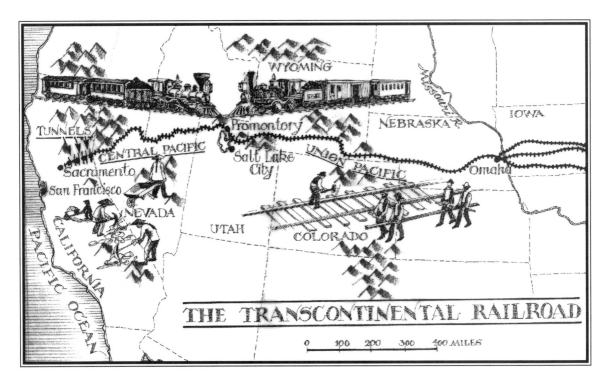

THE TRANSCONTINENTAL RAILROAD

0   100   200   300   400 MILES

A private Pullman car. The rich could now travel in this style clear across the country.

past in his private railroad car. When the transcontinental railroad was finished, Stanford, Durant, and the other railroad tycoons were national heroes. Today, many historians call them crooks.

Leland Stanford

You see, they had asked for government aid in building the railroad. That was reasonable. It was too big an undertaking for individuals. They demanded more than money from the government; they wanted—and got—enormous and valuable land grants. That was greedy, but it wasn't illegal. Then they sold stock in their companies to the public, and got more money that way. That wasn't illegal, either. But when their companies gave out contracts for building the railroads, and those men in charge —Stanford, Durant, and company—took all the contracts for themselves, that was crooked. They didn't even share profits with their stockholders. That was really foul play. Worse than that, they charged the government twice what it actually ly cost to do the building. As you know, they hardly paid their workers. And, still worse, they allowed poor and unsafe

62

Grant himself (in straw hat and beard, hands on fence) came to settle a fight between Dodge the engineer (far left) and his railroad owner, Durant (in straw hat, to right of top-hatted man). Durant wanted Dodge to make the railroad even longer than it had to be so he could get more government subsidies. Dodge refused, and Grant backed him.

workmanship, because it was cheaper and made their profits higher.

So that side of it was a mess, and when some of the story came out a few years later it created a big scandal. That was when U. S. Grant was president. It almost destroyed his term in office. All the fuss did help get laws passed to regulate business greed.

Still, to be fair, it took great imagination and some risk to finance the railroads. The men who did it had foresight and courage (even if they had no consideration for their employees, stockholders, or fellow citizens).

The good part of the story is that people in the United States could now travel from coast to coast in 10 days. Some people still went west in covered wagons, but it wouldn't be long before the wagons were history. Soon there were several transcontinental railroads. They were hauling things as well as people.

Railroads made the United States into a united country. People from different parts of the nation got to know each other. New immigrants could go west and settle. People on the West Coast could meet relatives back East. And, even if they didn't travel themselves, Americans could now buy goods from other states, or talk to people who had traveled about. The railroad made the country seem smaller.

What was it the Engines said,
Pilots touching,—head to head
Facing on the single track,
Half a world behind each back?
This is what the Engines said,
Unreported and unread.

. . .Said the Engine from the East:
"They who work best talk the least,
S'pose you whistle down your brakes;
What you've done is no great shakes,—
Pretty fair,—but let our meeting
Be a different kind of greeting.
Let these folks with champagne stuffing,
Not their Engines, do the puffing."

. . .Said the Western Engine, "Phew!"
And a long, low whistle blew,
"Come, now really that's the oddest
Talk for one so very modest.
You brag of your East! You do?
Why, I bring the East to you!
All the Orient, all Cathay,
Find through me the shortest way;
And the sun you follow here
Rises in my hemisphere.
Really, —if one must be rude,—
Length, my friend, ain't longitude."

—BRET HARTE

63

14 Taking the Train

Just like travelers today, early train riders who didn't want to pay extra for a bed spent the night seeking the least uncomfortable position.

How about taking a train across the country? If you've read Meriwether Lewis's journal (and in 1870 lots of people had) then you've already traveled across the West vicariously (vy-CARE-ee-us-lee), which means you've experienced it second-hand through another person's description. But now you can see the country for yourself. You can see the vast high-grass prairies, you can gape at the snow-topped Rockies, and you can marvel at nature-sculpted castles of stone in Utah. When you tunnel through Weber Canyon you'll see—standing proud and alone in a landscape of scrub—the Thousand-Mile Tree (named for its distance from Omaha). There is so much to see in this glorious country that of course you must go. Poet Walt Whitman says this:

> I see over my own continent the Pacific railroad
> surmounting every barrier,
> I see continual trains of cars winding along
> the Platte carrying freight and passengers.
> I hear the locomotives rushing and roaring,
> and the shrill steamwhistle,
> I hear the echoes reverberate through the
> grandest scenery in the world....
> Bridging the three or four thousand miles of land travel,
> Tying the Eastern to the Western sea.

Walt Whitman says "winding along the Platte." What and where is the Platte? See if you can find out.

Poets are writing about the transcontinental railroad as if it were the Northwest Passage that the European explorers sought. And in a way it is. The train will take you from the Eastern to the Western sea (which means from Atlantic to Pacific), and do it in only eight or ten days. Of

course, you will have to change trains in Chicago and again in Omaha and yet again in Ogden or Promontory, but that is all part of the adventure. When you get to the "Big Muddy," which is everyone's nickname for the Missouri River, you'll get out of the train and cross the river on the rickety ferry that goes between Council Bluffs and Omaha. (If you wait a few years a railroad bridge will be built across the river.)

The journey from Omaha to Sacramento will cost you about $40. That is, if you travel in the coaches everyone calls the *emigrant cars*. There you'll sit—night and day—on hard, wooden seats (without any padding at all). Some people bring straw cushions, and boards to stretch across the seats to make a place to lie down.

Don't worry, there is a stove to warm the car and an enclosed toilet and sink. The train will stop for meals, and you'll have about half an hour to bolt down whatever you find at the railroad station. Usually it is buffalo or antelope steak, fried potatoes, boiled corn, and coffee, which sounds pretty good, but, as one traveler said, "The chops are generally as tough as hanks of whipcord." And it is the same thing at almost every stop—for breakfast, lunch, and dinner. When William L. Humason of

Many of the emigrant-car passengers prepared their meals aboard the train rather than pay restaurant prices at the stations. One family on this train didn't want to leave the cat behind.

If your destination is Denver or Santa Fe, you can get off the train at Cheyenne, Wyoming, and take a stagecoach to those towns. Stages for Salt Lake City will be found at the stop at Deseret, Utah.

In 1872, a magazine wrote, "From Chicago to Omaha your train will carry a dining car. ...You eat...admirably cooked food, and pay a modest price."

These are probably the James Boys, with Frank sitting left and Jesse right. Let's hope they don't come after your train.

The cars on the very first railroads were pulled by horses, so it was natural to call a locomotive an *iron horse.*

Hartford, Connecticut, took the trip, he complained that the food was served in "miserable shanties, with tables dirty, and waiters not only dirty, but saucy." It was a different story in Evanston, Wyoming, where travelers praised the mountain trout dinners at the railroad stop.

If you can afford to pay $75, you'll ride "first class" in a coach car with padded seats and a bit more room, so you can have an easier time snuggling down at night.

But if you want to do this trip in style, for an extra $4 a night you can reserve a space in one of Mr. Pullman's Palace Cars. There you'll find fancy seats that convert into beds, other beds that fold down from near the ceiling, beautiful wood paneling, mirrors, reading lamps, carpeting, and attendants to fuss over you. Elegant restaurant cars serve meals on white tablecloths—the food is mighty good. Several Palace Cars have organs built inside and it isn't unusual for passengers to spend evenings singing or listening to the professional musicians who sometimes give performances. George Pullman claims he "invented railroad comfort," and most people agree.

(The railroads can't keep up with requests for the sleeping cars, although that makes the total cost of the trip about $100, which is more than the average American workingman earns in a month.)

When you board the train at Omaha, don't get scared if you hear stories of *wild Indians* and of buffalo charges. The railroad agent may tell you of those dangers because he happens to have an insurance policy to sell you, just in case.

If you're lucky you'll see a buffalo herd, and if you do the train may slow down, or even stop, so that you passengers can shoot your rifles through the window and try to kill some of the beasts—for the sport of it. (Partly because of this attitude, the buffalo won't be around long, so take a good look.) But you are even more likely to see antelope racing the train (they win) and villages of prairie dogs poking quizzical heads up out of the earth.

As to Indians, those you see will be traders, or settled on a reservation. The Indians who are still free stay as far as possible from the iron horses. Most of them are fighting for survival against the soldiers who have been sent West to "control" them. Real danger may come from storms. Rains and spring floods often wash out the track; blizzards have kept a few trains stuck in cold snowbanks for days; and tornadoes have lifted trains right off the track.

No, don't listen to what they say about Indians—but you do need to worry about train robbers, like Jesse James. The first time the James boys robbed a train (it was 1873), they loosened a piece of track and tied a rope to it. As soon as they heard a train whistling in the distance they pulled the rope. When the fast-moving locomotive hit the empty space it flew off the track and crashed onto its side. The engineer was killed and some passengers were injured. In the confusion, Jesse and his gang boarded the train, held up the passengers— that means they took their money and jewelry—got valuables from the baggage car, jumped on their horses, and headed for the hills.

But forget the danger and make the trip. You're getting an opportunity that won't last long. You may find it hard to believe, but in a few years this western land will be filled with houses and cities.

Passengers found it amusing to aim at buffalo from the train, but by the 1880s there were fewer and fewer herds left to shoot.

Pullman the Perfectionist

George Pullman's name became a synonym for luxury.

George Pullman wasn't like Thomas Durant or Leland Stanford. He was a gentleman who tried to be fair to his employees. He even built a model town for his workers with parks, a theater, a library, schools, and ball fields. It was an example of good town planning. Then he made rules for the town that he thought would make his employees healthy, happy, and well-behaved. That sounds fine, but it was the same mistake James Oglethorpe made when he founded Georgia.

He wanted a perfect colony, but you just can't make rules for other people. Most people want to govern themselves, and Americans expect to do so. So some of George Pullman's workers were unhappy. They called him *paternalistic* (which means that he acted as though being the person who paid his employees' wages gave him the right to tell them how to live their lives, too). Others called his ideas *welfare capitalism*. (See if you can figure out what that means.) Most of the workers agreed that Pullman's ideas were fine—but they just didn't like his making decisions for them. And they certainly didn't like it when he cut their wages. They went on strike. What happened is interesting and disturbing. You might want to read about it in an encyclopedia or a book about American labor, or in Book 8 of *A History of US*. It would be a great subject for a school report, because the Pullman strike got out of hand.

15 Fencing the Homestead

With a portable shack, a homesteader could establish one claim and then move on to another.

Imagine: you are settled on a farm in Kansas. You've worked hard, your crops are thriving, you're pleased with yourself. Then a cowpuncher decides to drive his herd to market and, though you're not right on the Chisholm Trail, you're near enough. The herd stampedes and longhorns trample your land. Wham, bang, squash! You have no crops. Nothing left of a year's work. And maybe no farm, because without a crop to sell, you don't have money to buy more seed and supplies.

That kind of thing happened. It was just one of the things that discouraged farming in the Plains states. The early pioneers and the forty-niners (who went west during the gold rush) leapfrogged over the plains and mountains and settled in the Far West. They called the plains the "Great American Desert" and believed it was no good for farming. They were wrong. The region would become one of the best agricultural areas the world has ever known. But they were right about one thing. It wasn't an easy place to be a farmer. There were hardly any trees and not enough water. The soil was wonderfully rich—the pioneers found native grasses tall enough to hide a man on horseback—but the weather was either blisteringly hot or frigidly cold, with tornadoes

The Plains states stretch from Texas to Canada and from Kansas and Iowa to the Rocky Mountains. The Plains states are Iowa, Kansas, Minnesota, Missouri, Nebraska, North Dakota, South Dakota, Wyoming.

"Any woman who can stand her own company ...and is willing to put in as much time at careful labor as she does over the washtub, will certainly succeed," wrote one homesteader.

thrown in just to keep people on their toes. Besides that, there were invasions of grasshoppers that ate crops, droughts that dried them up, and a loneliness on the open plains that drove some people mad.

None of that seemed to matter. The dream of many Americans was to have a farm. Land in the East was spoken for. Even with problems, those vast plains looked inviting to people who wanted land of their own. Winds blew much of the time, which meant that windmills could be used to pump water from deep wells. That water made irrigation possible.

In 1862 (which was during the Civil War), Congress passed a bill called the Homestead Act. It said that for $10 any citizen, or anyone who had filed papers to become a citizen, could have 160 acres of public land. That included women. As soon as the Civil War was over, a lot of people headed west to get land and become farmers. Some say a quarter of a million widows and single women were among those who became homesteaders.

Many homesteaders were immigrants—right off the boat. Some western settlements became all German, or all Danish, or all Swedish, or all Norwegian. Many immigrants tried to hang on to their original culture. Food was one way to do it. Greek, Polish, German, or Italian

The advertisements for land in Kansas and Nebraska never said what a cornfield looked like after a grasshopper plague had finished with it.

The Plains States are the heart of our nation, and that heart beats slow and sure year after year....Nowhere can we find a closer correlation of landscape and character than in the Plains States. The people there are, for the most part, as plain and level and unadorned as the scenery.
—WILLIAM INGE, PLAYWRIGHT

Successful or not, a pioneer had to get used to being a long way from anywhere.

The story continues on page 72.

69

America's frontier is moving West so quickly that some people become alarmed. Congress decides to save some wilderness from settlement and, in 1872, establishes our first national park—Yellowstone. It becomes the largest wildlife preserve in the country.

Hear the wind
Blow through the
buffalo-grass,
Blow over wild-grape
and brier.
This was frontier, and
this,
And this, your house,
was frontier.
There were footprints
upon the hill
And men lie buried under,
Tamers of earth and
rivers.
They died at the end
of labor,
Forgotten is the name.
—STEPHEN VINCENT BENÉT,
WESTERN STAR

Recess for the children of Pine Creek School, Livingston, Montana, and their teacher, Miss Sherman, in 1888,

Pronghorns Abounding

Brewster Higley, a Pennsylvania doctor, packed his bags and headed west to become a homesteader in Kansas. He was so happy in his new home that he wrote a poem about it called "The Western Home." A neighbor set the poem to music and gave it a new name, "Home on the Range." Before long everyone was singing it. When Higley used the words *buffalo* and *antelope* everyone knew what he meant, but the proper names for those animals are *bison* and *pronghorns*.

Oh, give me a home,
Where the buffalo roam,
Where the deer and the
antelope play,
Where seldom is heard
A discouraging word,
And the skies are not
cloudy all day.

Home, home on the range,
Where the deer and the
antelope play,
Where seldom is heard
A discouraging word,
And the skies are not
cloudy all day.

Home on the Grange

In 1867, Oliver Hudson Kelley founded a social and political organization for farmers called the National Grange of the Patrons of Husbandry. It grew rapidly, especially in Minnesota, Wisconsin, Illinois, and Iowa. The Grange was a way for farmers to band together and protect their interests. Working people were joining unions; farmers joined the Grange.

The grangers (that's what members were called) influenced lawmakers and established cooperative stores and mills. They made politicians pay attention to the farmers' concerns. Do you know what that word *husbandry* means? In medieval times a *husband* was the peasant who farmed his own land, the man of the family who provided for his household. He had to look after his crops and animals and use them economically, and from that we get one meaning of *husband*, which is "to be thrifty with one's resources." A meaning for *husbandry* that developed from this was simply "farming." From that came the meaning that the word usually has today: the application of scientific principles to farming, especially animal breeding.

And the word *grange*? It comes from England, where a grange was a farm or a farm building for storing grain, like a barn.

The Grange wakes sleeping farmers up to railroads' unfair practices—they made farmers pay more to ship their goods than the middlemen in cities did.

food was found in surprising places.

The newcomers didn't have to worry about buffalo herds anymore—the buffalo were practically gone—but they did have that problem of cattle wandering about. They couldn't fence their land because there were no trees to make fenceposts; besides, wooden fences rotted, or got knocked over, or burned. Joseph Glidden solved their problem. He invented barbed wire. He experimented in his backyard with an old coffee mill and a big grindstone that turned. He used them to twist two wires together, and then he coiled sharp barbs around the wires. With barbed wire, farmers could fence in their property.

Now the cowboys had a problem. Those fences got in the way of their herds. Well, cowboys and farmers did some fighting, but before long the

Early types of barbed wire. In 1874 Joseph Glidden put out 10,000 pounds of barbed wire.

farmers and ranchers won. The cattle drives were over and most cowboys turned into ranch hands. By 1890, railroads seemed to be about everywhere, so the cattle drives weren't necessary anyway. Railroads meant farmers and ranchers could send their cattle and grains to faraway markets.

A new kind of agriculture developed on the Plains. The early American farms had been self-sufficient. The farmer was able to take care of most of his own needs. Farm families raised cows, hogs, and chickens, grew wheat and vegetables, killed game, caught fish, built their own homes, and made their own furniture and clothes. They didn't have much use for money; they bartered for the few things they needed.

Self-sufficient farming wasn't suited to the Plains area or to the times. In the 19th century, agriculture became a big business. Many farmers became specialists who grew only one or two crops. It happened quickly. For thousands and thousands of years men and women had used the same methods of sowing and harvesting. Then a few inventions came along and changed everything.

Plains Writing

Willa Cather

July came on with that breathless, brilliant heat which makes the plains of Kansas and Nebraska the best corn country in the world. It seemed as if we could hear the corn growing in the night; under the stars one caught a faint crackling in the dewy, heavy-odoured cornfields where the feathered stalks stood so juicy and green. If all the great plain from the Missouri to the Rocky Mountains had been under glass, and the heat regulated by a thermometer, it could not have been better for the yellow tassels that were ripening and fertilizing the silk day by day. The cornfields were far apart in those times, with miles of wild grazing land between. It took a clear, meditative eye like my grandfather's to foresee that they would enlarge and multiply until they would be, not the Shimerdas' cornfields, or Mr. Bushy's, but the world's cornfields; that their yield would be one of the great economic facts, like the wheat crop of Russia, which underlie all the activities of men, in peace or war.

When spring came, after that hard winter, one could not get enough of the nimble air. Every morning I wakened with a fresh consciousness that winter was over. There were none of the signs of spring for which I used to watch in Virginia, no budding woods or blooming gardens. There was only—spring itself, the throb of it, the light restlessness, the vital essence of it everywhere; in the sky, in the swift clouds, in the pale sunshine, and in the warm, high wind —rising suddenly, sinking suddenly, impulsive and playful like a big puppy that pawed you and then lay down to be petted. If I had been tossed down blindfold on that red prairie, I should have known that it was spring.

Everywhere now there was the smell of burning grass. Our neighbors burned off their pasture before the new grass made a start, so that the fresh growth would not be mixed with the dead stand of last year. Those light, swift fires, running about the country, seemed a part of the same kindling that was in the air.

—WILLA CATHER, *MY ANTONIA*, 1918

On the Lone Prairie

The first Europeans in the New World were surprised by America's forests—by their vastness and vigor—but they weren't surprised by forests. Europe was full of trees. It was when the people pushed west, into the continent's heartland, that they found something that was indeed a new world. America's savannas—its grasslands—seemed endless. They were like nothing any of them had seen before. The grass sometimes reached 12 feet, so that the tallest animals and men were hidden in the growth. But if you stood on a rise and looked over the grass, there was nothing to block your view—no mountains, no trees—nothing. Just an enormity of sky that stretched out in every direction and rubbed its belly on the grass.

> The unshorn fields, boundless
> and beautiful,
> For which the speech of
> England has no name—

That was poet William Cullen Bryant's explanation for that French word: *prairie*. It means "big meadow." (There were savannas in Africa—with lions and giraffes and rhinoceros—but the African grasslands were much smaller than the American prairie.)

Our prairie divided itself into three regions. The *tallgrass prairie* began below Lake Michigan, in Illinois, and pushed west. (All of Iowa was filled with tall grass.) The tall grasses thrived where there was plenty of rain; sometimes those grasses—especially big bluestem—grew half an inch a day.

Far to the west, in the shadow of the Rocky Mountains, there were short grasses—just a few inches high. This was the Great Plains region. It was high, flat, dusty-dry grassland—a *steppe*—cold in winter and hot in summer. The grasses that grew best on the Great Plains—buffalo grass and blue grama—were drought-tolerant.

In between the tall and short were—as you might guess—mixed and medium grasses. Altogether it was the greatest grassland on earth, and home to wildflowers, birds, insects, and animals in astonishing balance and abundance. Before the railroads, the homesteaders, and the cattle ranchers pushed west, there were perhaps 60 million bison and 50 million pronghorn, along with millions of wolves, deer, elk, and coyote—as well as grizzlies, bighorn sheep, cottontails, rattlesnakes, and perhaps five billion (yes, billion) prairie dogs. (Prairie dogs aren't dogs at all, they are burrowing members of the squirrel family.) But there were more earthworms and butterflies than prairie dogs. And as for birds and ducks, in migration season they sometimes filled the sky like a dark moving cloud that blocked the sun from the earth and stretched as far as anyone could see.

Prairie grass has thick roots that twist and tangle and intertwine with the earth. That root-hard soil made the sod that the settlers cut for their homes. At first it broke the homesteaders' plows—but steel plows mastered the sod. Prairie fires kept the grasslands treeless. The fires started naturally, from sparks of lightning, and they spread—like wildfire. The fires were useful; they cleared out the dead grasses and encouraged new shoots. But animals—or people—were sometimes faced with terrifying walls of flame higher than their heads.

Domestic animals (cattle and sheep) and farmers (who pulled up the grasses and planted food crops) changed the prairie from grassland to market basket. The fertile land where grass grew so vigorously became the richest agricultural region in the world. The vast prairies turned into corn and wheatfields, or cities, or grazing lands, or sometimes forests (when fires were fought). Today, the produce of this region feeds our nation and others, too.

That market basket reminds us the earth is a changing place. The cornfields are just the latest inhabitants of a region rich in environmental history. In the great sweep of time, the grasslands were newcomers. One hundred million years ago, mid-America was a tropical jungle, with lush forests and roaming dinosaurs. Then the climate changed, the dinosaurs disappeared (to return on TV screens), and grass took over.

And what of that grassland? Where can you see prairie today—real prairie, like Lewis and Clark saw? Hardly anywhere. Illinois, which once had 37 million acres of tallgrass prairie (and is known as the Prairie State) now has about 3,500 acres of it. There is some tallgrass prairie at Konza Preserve, near Manhattan, Kansas, and the Nature Conservancy has a tallgrass preserve in Oklahoma, 17 miles north of Pawhuska. For midgrass, visit the Willa Cather Prairie near Red Cloud, Nebraska. You can see shortgrass prairie at Coronado National Grassland in Kansas. You'll also find prairie at Blue Mounds State Park in Minnesota and prairie-dog towns in Shirley Basin, Wyoming. To see prairie, along with an awesome cave, visit Wind Cave National Park in South Dakota.

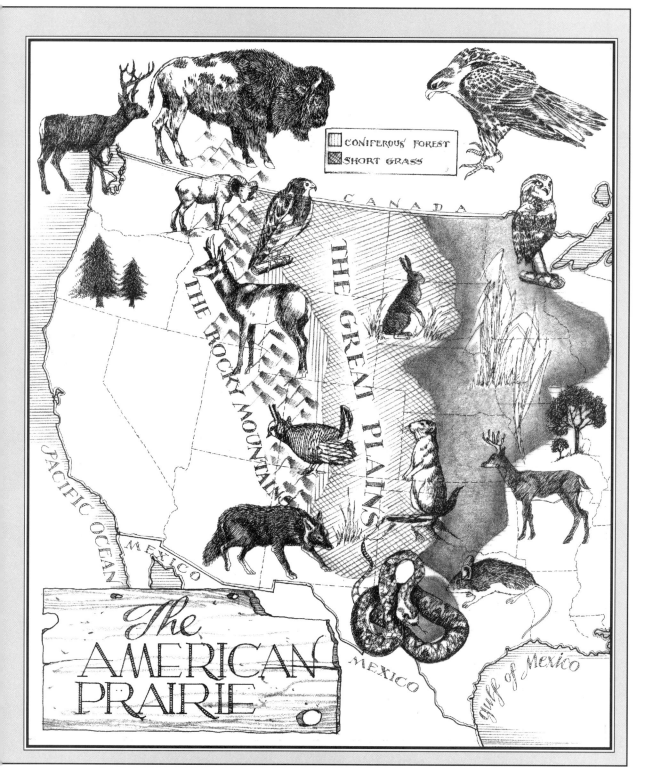

CONIFEROUS FOREST
SHORT GRASS

CANADA

THE GREAT PLAINS

THE ROCKY MOUNTAINS

PACIFIC OCEAN

MEXICO

MEXICO

Gulf of Mexico

The AMERICAN PRAIRIE

16 Reaping a Harvest

Cyrus McCormick was a Southerner who backed slavery. He thought Lincoln's election a disaster.

Sod is turf—grass with the soil that it grows on.

During the Civil War years Cyrus McCormick lived in England. He was dour, which means not much fun, and a big contributor to the Presbyterian church.

Plains land is fertile—you just learned that—but it is also firm, so firm you can dig it up in blocks, called sod blocks, and build a house with them. Try plowing that hard land. The old-style wooden plows broke. Iron plows didn't work either; soil stuck to the iron. John Deere designed a steel plow. It was strong and the soil fell away from it. It revolutionized agriculture.

But that was nothing compared to what Cyrus McCormick's reaper did. More than anything else, it was that reaper (and the railroads) that brought people west and changed the way they farmed.

A reaper cuts and harvests grain. Old-style farmers cut grain with a scythe, which is a hand tool. That was hard, back-breaking work. It took a good worker all day to reap an acre. If he had 40 acres to cut, that meant 40 days. In 40 days a crop would be overripe and rotten. McCormick's machine—pulled by horses or mules—could harvest a huge field in no time at all. The mechanical reaper did to wheat farming what Eli Whitney's cotton gin did to cotton growing. It made big farms practical.

Cyrus Hall McCormick was a Virginia boy, born in the beautiful Shenandoah Valley, of Scotch-Irish Presbyterian parents. (The Scotch-Irish believed in hard work. McCormick rarely wasted time or played games.) It was his father, Robert, who understood the need for a mechanical reaper and attempted to invent one. He mounted scissors-like knives on a long bar pulled at the side of a horse or mule or ox. Cyrus improved the invention.

He sold a few of his reapers in Virginia, but Shenandoah land is hilly and farms are small. When he took a trip to the Middle West—and saw the vast, flat plains—Cyrus McCormick knew his future was there.

Then he met the mayor of Chicago, who was a shrewd businessman; the mayor lent Cyrus money to build a big factory to make reapers. That factory became one of America's greatest business successes.

Before Cyrus McCormick came along, the Industrial Revolution had been mostly a city phenomenon (fuh-NOM-uh-non). McCormick brought that revolution to farm life. He was more than just a fine inventor. He was a business and marketing genius, too. He guaranteed his machines: if they broke down he saw that they were repaired—no one had done that before. He trained experts to show farmers how to use them; no one had done that before either.

Railroads brought large-scale commercial farming to the Midwest, and that began the reign of King Wheat. Wheat exports rose from 2 million bushels in 1860 to 90 million bushels in 1890. Minneapolis became the flour-milling center of the U.S.

A McCormick reaper pulled by a team of 32 first-class horses makes short work of a field in eastern Oregon. Smaller farms needed far fewer animals.

Bye, Bye, Blackbird!

In New England they once thought Blackbirds useless and mischievous to their corn, they made [laws] to destroy them, the consequence was, the Blackbirds were diminished but a kind of Worms which devoured their Grass, and which the Blackbirds had been used to feed on encreased prodigiously; Their finding their Loss in Grass much greater than their savings in corn they wished again for their Blackbirds.

Benjamin Franklin said that, but it took a long time for most Americans to realize that nature seems to know best.

The West was no place for an aristocrat. Any person who felt himself better than another quickly lost the respect of the community. There was no title nor term of respectful address. All were free and equal.

—EVERETT DICK, HISTORIAN

The machines were expensive, more than most farmers could afford. Cyrus McCormick let farmers take several months to pay for them. It was called installment buying—that was another of McCormick's ideas. He added a research department to his factory, and he kept improving the models. That, too, was something new for a businessman to do.

John Deere and Cyrus McCormick had developed the plow and reaper before the Civil War—in the 1830s and '40s. Their inventions were first widely used in the cleared woodland farms of the region to the east of the Mississippi. But after the war, those mechanical tools went west with all the people who were turning the American plains into farmland.

Note this statistic: in the 30 years between 1860 and 1890, more land was turned into farmland in the United States than in all the years from 1607 to 1860. In 1879 the McCormick factory produced 18,760 reapers; two years later it made nearly 49,000 machines. And it kept growing.

Farming was becoming an industry. The new equipment made huge, businesslike farms common. Farm equipment became necessary. Capital was now an important part of farming.

The farming revolution was hard on some people. Revolutions usually are. The small farmer was often hurt. More and more small farmers

By 1880, Cyrus McCormick's factory in Chicago produced over 100 machines a day. McCormick built his first factory in 1847, when Chicago was a town of 17,000 people. He guessed that railroads would make it important.

began heading for cities to take jobs in manufacturing and industry. Many didn't want to do that. They had no choice. They either became big or failed.

By 1900 the lone, self-sufficient farmer whom Jefferson admired hardly existed. The new farmer was part of a huge system. His wheat, cotton, beef, and wool were sold around the world. He had to worry about markets and prices in London and Chicago instead of in his neighborhood. He was not self-sufficient.

Because the country seemed so large, American farmers had always farmed wastefully. When land wore out, they just moved on to better land. For a long time there was better land to move to. But, by the end of the 19th century, there wasn't any frontier left. In addition, poor farming methods had destroyed more than 100 million acres of America's land. Prairie grasses and trees had been cut and plowed under, and then there was nothing to hold the soil in place: much good land turned to dust and blew away in the wind. Rain took the topsoil—the fertile part of the land—and washed it into streams. Land that had once supported buffalo and other wildlife became barren.

A few people were alarmed. Congress passed the Morrill Act. It gave the states large land grants to establish agricultural colleges. At those colleges farmers could learn the best and newest methods of farming. The Hatch Act established agricultural experiment stations in each state. Slowly the American farmer began to turn scientific. Mark Alfred Carleton went to Russia and found a variety of wheat that could take the tough weather in the American Middle West. Luther Burbank, on his farm in Santa Rosa, California, experimented and developed hundreds of varieties of new fruits, vegetables, grains, cacti, and flowers. And George Washington Carver, at the Tuskegee Institute in Alabama, developed hundreds of ways to use the peanut, the sweet potato, and the soybean.

> ## A New Nation Wows 'Em
>
> At first the new nation used imported brains —like Sam Slater's—to get its industry going. In 1851, at a great fair of industry held in a glass building in London called the Crystal Palace, people made fun of the American exhibition, and especially of McCormick's reaper. But, when it cut a swathe of wheat 74 yards long in 70 seconds, the laughing stopped. It had "mowed down the British prejudice," said a reporter. In the end, the reaper drew more visitors than the famous Koh-i-noor diamond. Europe now found itself turning with amazement to the New World for ideas and inventions.

One hundred million acres is about the size of Ohio, North Carolina, Maryland and Illinois combined. Some experts say twice that amount of land was seriously eroded.

Standing third from the left is George Washington Carver himself, teaching a class at Tuskegee. Check the blackboard for the subject being studied.

17 The Trail Ends on a Reservation

This is thought to be the only known photograph of Crazy Horse, the Sioux leader who beat Custer.

Misunderstandings—that's how trouble begins. Even Abraham Lincoln—who tried so hard to be fair—misunderstood the Native Americans. "We are not as a race, so much disposed to fight and kill one another as our red brethren [are]," said Lincoln. Read that again! Be sure you understand what Lincoln was saying. He made that statement during the bloody Civil War. It was one time he was *not* on target.

What Lincoln was doing is what many of us seem to do: he was lumping a lot of people together and making a general statement. But Indians are not all alike: some have warring traditions, others peaceful ones. Sioux are as different from Pueblo people as Swedes are from Turks.

That was part of the problem. The new Americans and the Native Americans were sharing the same land, but they didn't really know each other. They had different ways of living. Neither group wanted to change—and why should they?

The problem was that the two ways of life were not *compatible*. That means they couldn't exist together on the same land. And they both wanted that land. The Plains Indians were mostly hunters. The new settlers were mostly farmers and ranchers. Hunters and farmers have a hard time living together. Hunters need land free and uncultivated so herds of buffalo and deer and antelope can move about. Farmers need land cleared of wild animals so their crops won't be trampled, eaten, and destroyed.

When Lewis and Clark explored the West (when was that?), vast herds of buffalo stretched as far as the eye could see. By 1865 there were still

Some people have the idea that the Native Americans lived in perfect harmony with the landscape and never did anything to harm it. That isn't so. Some Indian peoples practiced environmentally sound farming methods, and others did not.

about 12 million buffalo. One observer told of a herd, moving at about 15 miles an hour, that was so big it took five days for the whole herd to pass him by. A few years later, the buffalo were just about all gone. They were hunted almost to extinction. Hunters like Buffalo Bill Cody led the charge, leaving herds where they fell. The land stank with the smell of dead buffalo. Then the plains, which had once vibrated with the sound of animal hoofs, became quiet as a desert. (But not for long; cattle soon replaced the buffalo.)

The abundance of the land had turned many Americans into wasters. There were so many buffalo, and so many trees, and so much land, it was hard to imagine an end to any of it.

Most Americans are individualists, and proud of it. But what is good for the individual farmer, or lumberman, or cattle rancher may

In 1905, this woman hanging meat to dry outside a tepee in Montana was still living much as her ancestors had.

As white hunters exterminated the buffalo, they destroyed the Plains Indians' means of support. "Your people make big talk, and sometimes make war, if an Indian kills a white man's ox to keep his wife and children from starving," said a Cheyenne chief. "What do you think my people ought to say when they see their buffalo killed by your race when you are not hungry?"

Picture 100 buffalo, then 1,000, then 10,000, then 100,000. One million buffalo are almost beyond imagining.

81

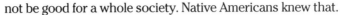

not be good for a whole society. Native Americans knew that.

The Indian cultures were usually centered on the community, not the individual. But what is good for the whole community may be frustrating for some individuals. These are two different ways of organizing people and governments. One way isn't better, or worse, than the other way. Each is just different, and worthy of respect.

The Indians of the Plains didn't get much respect. They depended on buffalo meat for food and buffalo skins for clothing and shelter, and they were horrified to see the buffalo and the land wasted. That left them hungry, angry, and confused. Those Native American men and women, who had once lived freely on the land, now faced machine guns, cannons, army troops, and the diseases the newcomers brought with them. They fought these new enemies with all the energy they had, but they didn't have a chance of winning.

The new Americans held power, so they got their way. All the Indians had was determination and courage. It wasn't long before the settlers had the land they wanted, and most of the Indians were dead or put on unwanted land called "reservations."

To better understand what happened, it helps to realize that after the Civil War there were many former soldiers who had learned, during the war, how to kill people. They were used to doing it, and they didn't know quite what else to do. So they came west and killed Indians. Peaceful settlers who moved west to farm were often innocent victims of angry Indians. All that led to intense hatreds and revenge raids on both sides. The Indians especially hated the iron horses (those roaring locomotives), which thundered across their hunting grounds filled with gun-toting soldiers and settlers.

A Sioux named Red Horse drew this picture of the batle of the Little Bighorn. One Sioux recalled how a cavalryman grabbed his braids and tried to bite off his nose.

The Native Americans weren't willing to give up their land and move onto the reservations, which were usually on poor land that was not good for growing crops. Would you be willing? They fought for their land and their way of life. Army troops—sent west to "control" Indians—slaughtered them instead.

"The only good Indian is a dead Indian," said Philip Sheridan, who had been a Union general during the Civil War. His boss,

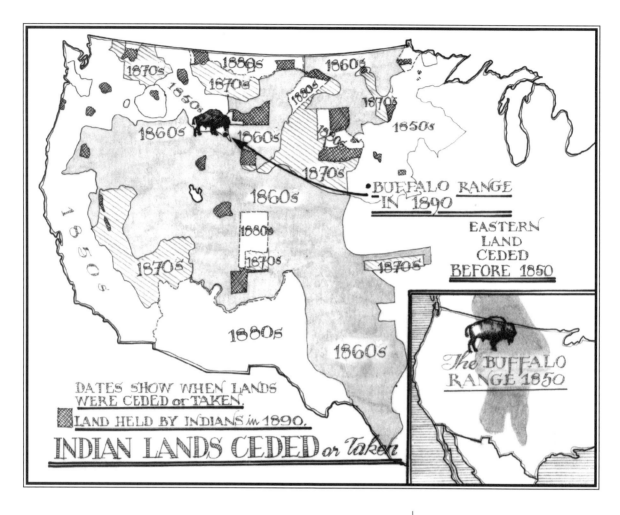

Buffalo Range in 1890
Eastern Land Ceded Before 1850
Dates Show When Lands Were Ceded or Taken.
Land Held by Indians in 1890.
INDIAN LANDS CEDED or Taken
The BUFFALO RANGE 1850

William Tecumseh Sherman, seemed to agree. Sherman's middle name was that of a great Indian hero, but Sherman didn't brag about that fact. He was sometimes called "Cump," but never *Tecumseh*. He talked about a "final solution" to the Indian problem. By that Sherman meant their destruction.

John Pope (who had been head of the Union forces at the second battle of Bull Run—and lost), announced that he would deal with the Sioux "as maniacs or wild beasts, and by no means as people with whom treaties or compromises can be made." He tried his best to do that.

In Colorado, Colonel John M. Chivington, a former minister, slaughtered 150 Cheyenne

John Pope

The more Indians we can kill this year, the less will have to be killed the next war, for the more I see of these Indians, the more convinced I am that they all have to be killed or be maintained as a species of paupers [poor people with no means of support whatever].

—GENERAL W. T. SHERMAN, 1867

83

Colonel John Chivington

As far as we know, General Sherman was the first person to use the phrase "final solution" about destroying a whole race of people. In the 20th century the Nazis (an evil political/military party in Germany) used that same phrase in their attempt to kill all the Jewish people in Europe. You will read about this idea again in Book 9 of *A History of US*.

These Sioux boys were sent to the Indian academy in Carlisle, Pennsylvania, to be turned into "Americans." On the right is what they became.

who had gone to the governor for protection. Most were women and children. Chivington called it "an act of duty to ourselves and civilization."

About one fourth of the army's western troops were black. The Indians called them Buffalo Soldiers because some had curly hair, like the buffalo. The men liked the name and used a buffalo as their emblem. Blacks were determined to be equal partners with whites in the American way of life. The Indians felt differently. Most had no desire to be partners: they wanted to be left alone to pursue their own way of life. Here are some words from a Minnesota Indian chief:

> *The whites were always trying to make the Indians give up their life and live like white men—go to farming, work hard and do as they did—and the Indians did not then know how to do that, and did not want to....If the Indians had tried to make the whites live*

like them, the whites would have resisted, and it was the same way with many Indians. The Indians wanted to...go where they pleased and when they pleased; hunt game wherever they could find it, sell their furs to the traders, and live as they could.

The chief told how the Indians were herded onto reservations where the land was poor and there was not enough to eat. They had to buy food and supplies from government agents who usually cheated them.

Many of the white men often abused the Indians and treated them unkindly. Perhaps they had excuse, but the Indians did not think so. Many of the whites always seemed to say by their manner when they saw an Indian, "I am much better than you," and the Indians did not like this...the Dakota did not believe there were better men in the world than they.

The Indian wars in the West—the hardest fought of them—lasted from the end of the Civil War (1865) until a final massacre of Indians in 1890 at a place called Wounded Knee. Could the land have been shared? It wouldn't have been easy. Where do you live? Who lived on your land 300 years ago? Would you share your home? What about homeless people today? How can we solve their problems?

The Indian story should have been different. There could have been respect and honesty between the peoples. There could have been strong laws to prevent unfairness and brutality. There could have been more understanding. Even those—like the Christian missionaries—who meant to help the Native Americans usually ended up destroying the tribes because they didn't respect the native

In 1884, the Pine Ridge Indian Agency School in Dakota Territory (a few miles from Wounded Knee) looks prisonlike with its barbed-wire fence. Such schools tried to suppress Indian ways.

We had buffalo for food, and their hides for clothing and for our teepees. We preferred hunting to a life of idleness on the reservation, where we were driven against our will. At times we did not get enough to eat, and we were not allowed to leave the reservation to hunt. We preferred our own way of living. We were no expense to the government. All we wanted was peace and to be left alone. Soldiers were sent out in the winter, who destroyed our villages. Then Long Hair [Custer] came in the same way. They say we massacred him, but he would have done the same thing to us had we not defended ourselves and fought to the last.

—CRAZY HORSE, SIOUX LEADER, JUST BEFORE HE DIED (1877)

The story continues on page 88.

Pitiful Last Becomes a Winner

As a doctor, Charles Eastman treated survivors of the Wounded Knee massacre.

When the boy child was born he was given a name that, in the Sioux Indian language, means "Pitiful Last." That was because he was the last of his mother's five children. She died giving birth to him.

His father was Many Lightnings, but it was his grandmother who raised him.

Pitiful Last grew strong and quick and bright; when he was still a boy, but close to manhood, he won a fiercely fought lacrosse game. It was then that he received a new name. It was Ohiyesa; it means "The Winner."

But Ohiyesa was not through with names. He was raised as a Plains Indian, but his mother's father was a white man, an army officer and an artist. He was Seth Eastman. So the boy had another name: it was Charles Eastman.

It was well that he had two names, because he lived in two worlds. When he grew up he went to Dartmouth College, and to Boston University, and learned to be a doctor. Then he returned to the Indian world and served his people as a physician. But he wasn't through learning. He studied law and represented the Sioux nation as an attorney in Washington. Are you amazed that he was both a doctor and a lawyer? Well, that isn't all he was. He is perhaps best known as a writer.

One of the books he wrote is about his childhood. It is called *An Indian Boyhood*. Here is part of it:

"With the first March thaw the thoughts of the Indian women... turned promptly to the annual sugar-making," wrote Charles Eastman / Ohiyesa of maple-sugar time in Minnesota. Mostly it was the old men and women and the children who made maple sugar. "The rest of the tribe went out upon the spring fur-hunt...leaving us home to make sugar." The work began when

A MAPLE TREE WAS FELLED AND A LOG canoe hollowed out, into which the sap was to be gathered....My grandmother worked like a beaver in these days (or rather like a muskrat, as the Indians say; for this industrious little animal sometimes collects as many as six or eight bushels of edible roots for the winter, only to be robbed of his store by some of our people). If there was prospect of a good sugaring season, she

now made a second and third canoe to contain the sap. These canoes were afterward utilized by the hunters for their proper purpose....

A long fire was now made in the sugar house, and a row of brass kettles suspended over the blaze. The sap was collected by the women in tin or birchen buckets and poured into the canoes, from which the kettles were kept filled. The hearts of the boys beat high with pleasant anticipations when they heard the welcome hissing sound of the boiling sap! Each boy claimed one kettle for his especial charge. It was his duty to see that the fire was kept up under it, to watch lest it boil over, and finally, when the sap became syrup, to test it upon the snow, dipping it out with a wooden paddle. So frequent were these tests that for the first day or two we consumed nearly all that could be made; and it was not until the sweetness began to pall that my grandmother set herself in earnest to store up sugar for future use. She made it into cakes of various forms, in birchen molds, and sometimes in hollow canes or reeds, and the bills of ducks and geese....

Ohiyesa remembered:

I OFTEN FOLLOWED MY OLDER brothers into the woods, al-

though I was then but four or five years old. Upon one of these excursions they went so far that I ventured back alone. When within sight of our hut, I saw a chipmunk sitting upon a log, and uttering the sound he makes when he calls to his mate. How glorious it would be, I thought, if I could shoot him with my tiny bow and arrows! Stealthily and cautiously I approached, keeping my eyes upon the pretty little animal, and just as I was about to let fly my shaft, I heard a hissing noise at my feet. There lay a horrid snake, coiled and ready to spring! Forgetful that I was a warrior, I gave a loud scream and started backward; but soon recollecting myself, looked down with shame, although no one was near. However, I retreated to the inclined trunk of a fallen tree, and

Seth Eastman painted this picture of Sioux playing lacrosse in 1851, seven years before the birth of his lacrosse-playing grandson Charles / Ohiyesa.

there, as I have often been told, was overheard [saying]…"I wonder if a snake can climb a tree!"

The sugaring went on well into April, when birds, returning from their winter trip south, began singing in the camp.

From Jamestown to Wounded Knee

1607: first English settlers arrive in Virginia. No one knows the number of Indians on the continent; some estimates are as high as 100 million.

1607–1750: cooperation and encroachment. Efforts made to convert Indians to Christianity. Many deaths from disease, wars, enslavement.

1787: Northwest Ordinance recognizes existence of Native American property.

1824: Bureau of Indian Affairs formed within War Department.

1830: Indian Removal Act leads to Trail of Tears and other relocations of southeastern tribes.

1840s–'60s: Eastern tribes pushed continually westward; some, like the Sauk and Fox, resist strongly but are eventually overcome.

1871: Congress reverses the Northwest Ordinance. Tribes are no longer independent nations with whom treaties can be made.

1876: General George Armstrong Custer's troops destroyed by Sioux warriors under Sitting Bull at battle of Little Bighorn.

1887: Dawes Act dissolves tribes as legal entities that can own land.

1890 Sitting Bull shot and killed by Indian policemen.

1890: at least 150 Sioux warriors, women, and children massacred at Wounded Knee in the last major armed encounter between Indians and whites in North America.

1924: Indian Citizenship Act.

1934: the Indian Reorganization Act attempts to restore tribal structures.

"Let it be recorded," said Sitting Bull of the Sioux, "that I am the last man of my people to lay down my gun."

cultures. They were arrogant, and they didn't realize it. That was because they were sure their way of life was better than the Native American way. They thought they were doing right when they tried to force Indians to live as they did. What would you have done?

Most of the new Americans just didn't seem to care much about Indians. They talked of "conquering" the land and its ancient peoples. And that was what they did in brutal Indian wars. Should they have thought of cooperation rather than conquest? Do you think it would have worked? The past can't be changed, but can we learn from it? Can a modern, industrial country protect native peoples and the natural environment?

During the 19th century, the West of the traditional Indians, the mountain men, and the buffalo reached its end. A new and different West came into being: a land of farmers, ranchers, miners, city dwellers, and Indians who had to adapt their ways to new realities. But before the change was completed, before the final act in the drama, a great leader attempted to save his people. His name was Chief Joseph.

Straight Talk

Bishop Whipple

The Indians did have some friends. One was Henry Benjamin Whipple. They called him Straight Tongue and meant it as a compliment. They knew he would not lie to them. Whipple was Episcopal bishop of Minnesota. He got Abraham Lincoln to pardon a large number of Sioux who were unjustly sentenced to death. Some Minnesotans were furious when they heard what Whipple did. They wanted to get rid of all the Indians. It was pure selfishness: they wanted Indian land.

18 The People of the Pierced Noses

Before Chief Joseph's father died, he told his son: "Stop your ears whenever you are asked to sign a treaty selling your home."

The Nez Perce (nez-purse) Indians were special. Everyone agreed about that. They were honest and honorable, free-spirited and courageous, intelligent and independent, handsome and well-built, and, as if all that weren't enough, they lived in a region that was a kind of paradise. Their land—where today Idaho, Washington, and Oregon come together—holds lush valleys, grassy prairies, steep mountains, and canyons that seem to have been cut by a giant's steam shovel.

The Nez Perce shared that land with elk, deer, antelope, rabbits, fowl, and mountain goats (along with some predator enemies: bears, wolves, foxes, and coyotes). Fish, especially the lordly salmon, splashed in their streams. According to one observer:

> *from the first breath of spring until midsummer, the Nez Perce country is a blaze of color. Blue windflowers, purple shooting stars, yellow bells, bluebells, blue and purple penstemon, blue and yellow lupine, yellow sunflowers…and, above all, the camas, covering the open meadows with blue carpets until at a distance they resemble little lakes.*

The Nez Perce were mighty hunters, and known for their strong bows. Most were fashioned of cherrywood or yew, but the best of them were made of the horns of the mountain sheep, which were boiled and bent and backed with layers of sinew. Other tribes traded their most precious goods for those bows.

When horses arrived in this northern region, the Nez Perce quickly be-

If you want to learn more about the Indians of the West and about the Indian wars (and heroism, too), read about Crazy Horse, Sitting Bull, Black Hawk, and Geronimo.

Understand me fully with reference to my affection for the land. I never said the land was mine to do with it as I chose. The one who has the right to dispose of it is the one who has created it. I claim a right to live on my land, and accord you the privilege to live on yours.

—CHIEF JOSEPH

Like Chief Joseph, Crazy Horse led his men in a heroic fight against odds that could not be overcome. But at a place called Little Bighorn the Indians triumphed. Reckless George Armstrong Custer (who had won fame during the Civil War) and his troops were wiped out by a much larger Indian force. The victory didn't last for long.

The Indians of the Wild West
We found were hard to tame,
For they seemed really quite
*   possessed*
To keep their ways the same.
They liked to hunt, they liked to fight,
And (this I grieve to say)
They could not see the white man's
*   right*
To take their land away.
So there was fire upon the Plains,
And deeds of derring-do,
Where Sioux were bashing soldiers'
*   brains*
And soldiers bashing Sioux'.
And here is bold Chief Crazy Horse,
A warrior, keen and tried,
Who fought with fortitude and force
But on the losing side.
Where Custer fell, where Miles
*   pursued,*
He led his native sons,
And did his best, though it was
*   crude*
And lacked the Gatling guns.
It was his land. They were his men.
He cheered and led them on.
—The hunting ground is pasture,
*   now.*
The buffalo are gone.
                    —STEPHEN VINCENT BENÉT,
                    A BOOK OF AMERICANS, 1933

A headman of the Nez Perce and his wife. The Nez Perce were not part of mainstream Plains Indian culture, but they adopted many Plains customs such as the dog travois, which was used a lot before horses came.

came skilled riders, among the best in the land. Horses thrived on the high, abundant pastureland, and the Nez Perce learned to breed the animals for strength and beauty and fleetness.

They lived in a kind of democracy where individuals were respected. But their society wasn't perfect. The Nez Perce had enemies, and, although they loved peace, they fought frequently and captured slaves who then worked for them and had no say in their village affairs.

Even before they saw the first outsiders, they knew of them from other tribes. There were Spaniards to the south, Russians in Alaska and on the West Coast, and French to the north and east. It was the French who gave them their strange name. In French it is pronounced nay-pair-SAY, and it means *pierced nose*. The French had seen a few Indians with bits of clamshell decorating their noses. It was a fashion of some West Coast tribes, but not usually of these people.

The Nez Perce must have been surprised when Meriwether Lewis and William Clark stumbled into one of their camps. It was late September, in 1805, and the members of that expedition, sent to explore the West by President Thomas Jefferson, had been caught in a mountain snowstorm. They were starving.

I met 3 Indian boys, when they saw me [they] ran and hid themselves, wrote Clark (who wasn't much of a speller and who used punctuation poorly, too). Clark found the boys and *gave them Small pieces of ribin & sent them forward to the village Soon after a man Came out to meet me with great caution & Conducted me to a large Spacious Lodge.* The Indians fed Clark and his men buffalo steak and camas roots and probably saved their lives.

They liked each other—the Nez Perce and the explorers. Lewis and Clark convinced the Native Americans to stop the warring between tribes; that would make it safe for white men to open trading posts where they would sell guns, mirrors, and other goods. The Indians wanted those goods, and they held a council and promised "to cultivate peace."

It was the beginning of a real friendship. The Nez Perce rendezvoused (RON-day-vood) with mountain men, helped trappers and traders, and befriended those who were beginning to pass through their land. Since they were people who kept their word, they did "cultivate peace" with the newcomers.

Until gold was found on their land in 1860, that was easy to do. But then miners couldn't be kept off the land. Settlers followed. It was the beginning of troubles. Some Nez Perce signed treaties to give up some of their land, but others wouldn't do it. They wouldn't sign any treaties.

President Grant tried to solve the problem; he set aside a section of land "as a reservation for the roaming Nez Perce Indians." Settlers were not allowed on that land. But that didn't stop the miners and homesteaders. They defied the president. They moved onto the land. One of the "no-treaty" tribes was led by a man most Americans called Chief Joseph. His real name was *Hin-mah-too-yah-laht-ket*, which means "Thunder Rolling in the Mountains." Joseph told his people to be patient. He didn't want to fight the white settlers.

In 1876 (which was how long after the visit of Lewis and Clark?), the United States government sent three commissioners to meet with Chief Joseph. They wanted to persuade him to move from his land to another reservation. Joseph was thirty-six years old. "Straight and towering, he seemed strangely amicable and gentle; yet he bore himself with the quiet strength and dignity of one who stood in awe of no man," writes Indian expert Alvin M. Josephy, Jr.

No matter how the government commissioners pleaded, Joseph

Ollokot was Joseph's younger brother. They were very close; Ollokot was a mapmaker and a leader in battle.

The truth was that Nez Percé successes were resulting from a combination of overconfidence and mistakes on the part of the whites, the rugged terrain that made pursuit difficult, and, to a very great extent, the Indians' intense courage and patriotic determination to fight for their rights and protect their people.

—ALVIN M. JOSEPHY, JR.

When Chief Joseph surrendered, in 1877, the Indian wars were, for the most part, over. For the next 13 years there were minor skirmishes. Wounded Knee, in 1890, was the very end.

91

"We could have escaped from Bear Paw Mountain if we had left our wounded, old women, and children behind," said Chief Joseph. **"We were unwilling to do this. We had never heard of a wounded Indian recovering while in the hands of white men."**

would not agree to move. "We love the land," he said. "It is our home."

But the Nez Perce had no choice; the newcomers had great power. The Indians were to be forced onto a reservation. The commissioners had no patience. The Indians must go, and quickly, they said, even though the weather was bad. One frustrated, angry young Indian, whose father had been murdered by white settlers, killed some of the white men. Now the whites had a reason to call the Indians savages. Now they could attack.

When they were attacked, the Nez Perce fought. The first battle began when Indians, carrying a white flag of truce, approached the soldiers. A shot rang out and the Indians returned the fire. The fight, which was unexpected, was brief: 34 troopers died, and no Indians. The fleeing soldiers dropped their weapons—63 rifles and many pistols. It was a bonanza for the Indians. But they were few in number and they knew that an alarm would go out. Other soldiers would soon be after them.

So they raced for the place where they thought they would be free. They raced for Canada. It turned out to be a 1,000-mile journey. First one army, then another, and another, followed and fought them. Now Joseph proved to be strong as thunder. He led his small band brilliantly—although most were children and old people. They fought in their mountains, they fought in their valleys, they fought in their canyons, they fought on their plateaus. Everywhere they were outnumbered and outgunned. Over and again they outwitted their pursuers. But they were fighting the telegraph as well as an army. Fresh troops were summoned by wire. Finally, just 30 miles from Canada, facing new soldiers, the Nez Perce were surrounded.

The warrior Looking Glass advised staying in Montana rather than fleeing to Canada. He was killed by a U.S. army sharpshooter.

On their way toward Canada, the Nez Perce captured a stagecoach. They let the people inside escape, and then they had a good time riding in the painted vehicle—until cavalry appeared, and they had to jump out and fight.

Another time, they captured a group of sightseers in Yellowstone National Park. Joseph insisted that they be treated humanely. They were left behind when the Indians fled on.

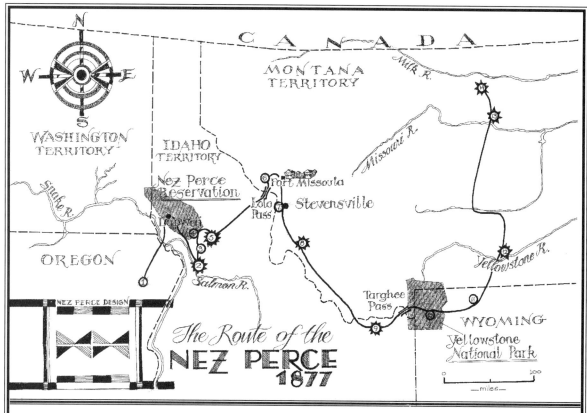

The Route of the
NEZ PERCE
1877

Happenings Along the Way

1. *June 12, Wallowa Valley. Four whites killed in a raid of revenge.*
2. *June 17, battle of White Bird Canyon. Killed: 34 soldiers, 0 Nez Perce.*
3. *Scouting party killed by Nez Perce.*
4. *Other Nez Perce join the retreat. Group totals about 550 women, children, and elderly, and 150 able men.*
5. *July 11, battle at Clearwater. Attack by 600 soldiers. Warriors fight while others escape. Killed: 13 soldiers, 4 Nez Perce. Band votes to escape to Montana.*
6. *Nez Perce bypass soldiers' barricades, later named Fort Fizzle.*
7. *Nez Perce peacefully trade for supplies.*
8. *August 9, battle of Big Hole. Surprise attack by 200 troops. Of 89 Nez Perce killed, 77 are women and children.*
9. *August 20, battle of Camas Creek. War party doubles back and attacks to gain time, cutting 200 army pack mules loose.*
10. *Vacationing tourists startled to see Nez Perce traveling through the new park.*
11. *Seek help from Crow but disappointed. Decide to escape to Canada.*
12. *September 13, at Canyon Creek, 350 troops catch up and battle Nez Perce. Again warriors hold them off while women and children escape.*
13. *September 25. Some Nez Perce raid army garrison for much-needed supplies.*
14. *September 30, battle of Bear Paw. Surprise attack by fresh troops leaves many Indians dead. Some escape to die, a few escape to Canada. When Chief Joseph surrenders, there are 350 women and children and 80 men.*

Emma and George Cowan were two of the Yellowstone Park tourists held for a while by the Nez Perce. George Cowan was left for dead and crawled 10 miles to the party's earlier camp because his legs were paralyzed—he had been wounded in the leg and rolled all the way down a hill.

Joseph ranks with Lee, Jackson and Grant as one of the best generals this country has produced.
—CHESTER A. FEE, BIOGRAPHER

Chief Joseph spoke. Here is what he said:

I am tired of fighting. Our chiefs are killed.…The old men are dead. …the little children are freezing to death. My people, some of them, have run away to the hills, and have no blankets, no food; no one knows where they are…my heart is sick and sad. From where the sun now stands I will fight no more forever.

That day, promises were made to Chief Joseph, but they were never kept. In Washington, the government people did not know the great chief and the brave Nez Perce. Those who wanted Indian land told false stories. The Nez Perce Indians were sent to an empty plain; most sickened and died. Chief Joseph pleaded for justice. Here is some of what he said:

All men were made by the same Great Spirit Chief. They are all brothers. The earth is the mother of all people, and all people should have equal rights upon it. You might as well expect the rivers to run backward as that any man who was born a free man should be contented when penned up and denied liberty to go where he pleases.

Are you reading this quickly? Well, don't do that. These are great words. They are worth rereading. They are worth memorizing.

We only ask an even chance to live as other men live. We ask to be recognized as men. We ask that the same law shall work alike on all men. If the Indian breaks the law, punish him by the law. If the white man breaks the law, punish him also. Let me be a free man—free to travel, free to stop, free to work, free to trade where I choose, free to choose my own teachers, free to follow the religion of my fathers, free to think and talk and act for myself—and I will obey every law, or submit to the penalty.

Here are more of his words:

Whenever the white man treats the Indian as they treat each other, then we will have no more wars. We shall all be alike—brother of one father and one mother, with one sky above us and one country around us, and one government for all.

The time would come, as Chief Joseph wished, when there was one government and equal rights for all men and women of every color and background. Chief Joseph's words would help bring that time. But it would be too late for most of the Nez Perce.

Chief Joseph and the officer who pursued him, General Howard, met once more in 1904.

19 A Villain, a Dreamer, a Cartoonist

Artist Thomas Nast's original caption for this Boss Tweed cartoon read: "Well, what are you going to do about it?"

Do you ever worry about air pollution or about dishonest politicians? Well, so did people in the 19th century.

Just to reassure you—most politicians are honest and most air is clean. But that is no reason to relax. There are people around who will mess up the world if we let them.

One of the worst, in the years after the Civil War, was a man named William Marcy Tweed. He was called "Boss" Tweed and he ran New York City. New York had problems—big problems—especially problems of air pollution and traffic congestion. Some 700,000 people lived in New York, most of them squeezed into a small area near the tip of Manhattan island. Much of the city's business took place around a famous street called Broadway. Trying to walk or take a horsedrawn bus down Broadway was a nightmare. There were so many people it sometimes took an hour just to move a few yards. And talk about pollution—whew—hold your nose while I tell you about it.

New York was home to more than 100,000 horses. Now, a healthy horse dumps between 20 and 25 pounds of manure a day. Imagine all that smelly manure spread around by wheels and feet. When the manure dried it turned into powder that blew in your face and went up your nostrils. But that wasn't the worst of it. In the 19th century, people and businesses could burn anything they wanted. Mostly that was

Who's the Boss?

William Marcy Tweed was called "Boss" Tweed because that was exactly what he was: the boss. He wasn't elected to run New York, but he did it anyway. He was actually a city alderman. (An alderman is a member of a city legislative body.) He was never mayor. It didn't matter. He controlled the New York State Democratic Party and the Tammany Hall political machine.

Tweed put graft (getting dishonest money) on a businesslike basis. All city contracts were padded by a fixed amount, which went to Tweed and his cronies. At the Tammany clubhouse, he slept in a bed with blue silk sheets. He sometimes entertained on his yacht, the *William M. Tweed*, which had a crew of 12, fancy furniture, and Oriental rugs. When Tweed went to jail, he was asked his occupation. He said, "Statesman."

A traffic jam on West Street, near the Hudson River docks where goods were loaded and unloaded. At New York's meat market, which is still on West Street, you can see the same cobbled streets built for these 19th-century wagons. Below, snow removal under Tammany.

coal, which puts black fumes in the air. Even worse, Standard Oil had a New York refinery. Oil refineries, without controls, give off terrible, noxious fumes. That oil refinery was a big polluter. Hold on, that's not all. When Boss Tweed controlled New York there wasn't much in the way of sanitary services. So people often dumped their garbage in the streets. Garbage smells—especially in August. Are you choking? Well, I still haven't mentioned the pigs. Pigs ran about eating garbage and leaving their own smells and dumplings. And then there were flies, and disease. But you may have heard enough.

There you are in the middle of Broadway, and you want to get away. You climb on a horsedrawn bus. It sways back and forth so violently that some passengers get seasick. You try walking. But there are no street lights (they haven't been invented yet). Horses, people, buses, and carriages are all pushing and shoving on Broadway. Pedestrians often get killed in traffic accidents. Have you had enough of the good old days? So had a lot of people in the 19th century. The politicians said that soot in the air was a sign of modern progress, but most people were beginning to gasp for fresh air.

Fresh air was the last thing that Boss Tweed cared about. He was a scoundrel—a real bad guy who controlled most of the city's jobs and services. He used his power to get money for himself. He bribed others and then forced them to do what he wished. Here is an example of the way he worked. A new city building was to be built; Boss Tweed became the contractor and charged the city three or four times what the building actually cost. He put the difference in his pocket. Then he filled the building with $50 sofas and charged the city $5,000 for each of them. How did he get away with that? Well, he was charming—in a scoundrelly way—so he fooled people. Many citizens didn't realize that he was stealing from them. And because he was so powerful, those who did know were afraid to do anything about it.

Except for a quiet, frail little man named Alfred Ely Beach. Beach was a genius—an inventor, a publisher, and a patent lawyer. He invented one of the world's first typewriters. He called it a "literary piano." Beach invented other things, too, and because of that he understood about patents. If you invent something, and you want to be sure that your idea is not stolen, it is necessary to register your idea with the patent office in Washington. As a patent lawyer, Beach helped many inventors with their patents.

Beach did many things and did them all well. When he was

Machinery of Government

"Who took the money?" is the question. "HE DID!" says everybody. Boss Tweed (the fat man with the huge nose) made the city pay $1.8 million for plastering one building—guess who the surplus came back to?

There were some things the Founding Fathers hadn't expected. Political parties were one. Political machines were another. The machines were unofficial governments that existed alongside the real city governments; each had its own functions. In New York, Tammany was the stronger one. Tammany Hall was the most powerful of all the urban political machines (machines thrived in many 19th-century cities). Insiders had their own name for Tammany; it was "the Tiger." That beast did some good—actually quite a bit of good—especially for immigrants who needed help getting started in America—but it was all at a price.

Tammany began as a kind of benevolent club in 1789, designed to help the poor. Mostly it was a dress-up society where men donned Indian garb, called themselves "sachems" or "braves," held parades, and drank whiskey. But when one of its early leaders, William Mooney, stole $4,000, he set the Tiger to thieving. Tammany paid people to vote—sometimes a dozen times each. And bribery? Well, the Tiger bribed the police, the elected officials, anyone who would take money. But eventually, when the immigrant flow slowed down and reformers closed in, the Tiger lost its roar. Do you hear people complain about politics today? Tell them to study some history.

only 19 he took over a small magazine named *Scientific American* and helped make it the fine journal it is today. He became publisher of the *New York Sun* and it became an important newspaper.

But that isn't what this story is about. It is about Ely Beach's fight

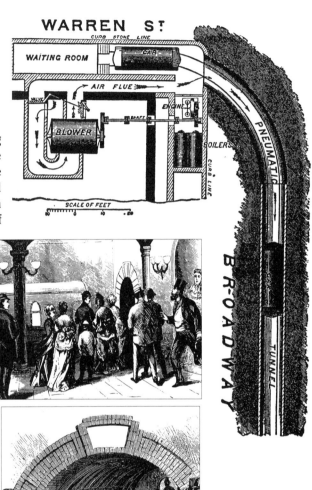

The subway's blower sucked air in from a grating on the street. If you were standing near the grating when the fan sucked, your hat came off; if you were there when it blew, all the street garbage hit your ankles. After each ride, daring passengers were allowed to stroll along the tunnel.

with Boss Tweed. Beach wanted to do something about the traffic congestion on New York's streets. He thought and thought and finally came up with the idea of putting a railroad train underground. He called it a subway. He knew Boss Tweed wouldn't let him build it—unless he agreed to give Tweed millions of dollars. And Beach was too honest a man to pay off a politician.

He decided to build a subway and not tell Tweed. He built it right under Broadway and hardly anyone knew he was doing it! He invented a hydraulic tunneling machine and a pneumatic subway. He got laborers to work at night and to haul dirt away in wagons with wheels muffled so they wouldn't make noise. It took 58 nights of secret work to get the tunnel done.

In February 1870, a group of New York newspaper reporters got invited to a reception. They were surprised when they were led underground into a beautiful, large waiting room. Paintings hung on bright walls, a pianist played at a grand piano, a fountain splashed, and goldfish swam in a giant tank. Beach had done it! His subway was ready. The reporters all took a ride in a cylinder-shaped wooden car. The car had handsome upholstered seats, fine woodwork, and gas lamps. It fit tight—like a bullet in a rifle—and moved down tracks inside a round brick underground tube. It went right under Broadway, under all the pollution and traffic.

What made it move? A giant fan blew it 371 feet. There the subway car stopped and tripped a wire; that made the fan reverse itself, and that sucked the subway car back.

Beach saw his subway as a model for a grand subway he had planned. It would carry 20,000 passengers a day and go for five

Hydraulic means water-powered; ***pneumatic*** means air-powered.

miles—to Central Park, he said—at a speed of a mile a minute. A mile a minute? Nothing had ever gone that fast.

Boss Tweed was outraged! He controlled all the streetcars in the city. This was a threat to his power. He must have pounded his diamond-ringed fingers. He got in touch with the governor—*his* governor (he'd bribed and bought him).

What happened is a long and complicated story, and I can't tell it all here. People flocked to Beach's little subway; they rode back and forth under Broadway. Beach gave the subway's profits to charity. The state legislature passed a bill allowing Beach to build the grand subway. Tweed's governor vetoed it. Beach worked hard, talking to congressmen, and a second subway bill was passed. Governor John T. Hoffman vetoed it again.

Finally, the newspapers began writing editorials telling the truth about Boss Tweed. A cartoonist—named Thomas Nast—drew funny cartoons that showed Tweed as the wicked man that he was. Tweed threatened Nast. "I don't care what the papers write about me—my

Fighting cartoonist Thomas Nast's self-portrait.

George Washington *What?*

George Washington Plunkitt: his name sounds unreal, and some of what he said seems so too, but he was real, and a member of Tammany Hall, and a New York state senator. Journalist William Riordan, who knew Plunkitt, called him "Tammany's philosopher," and took down some of his words.

Everybody is talkin' these days about Tammany men growin' rich on graft, but nobody thinks of drawin' the distinction between honest graft and dishonest graft," said Plunkitt. "There's all the difference in the world between the two. Yes, many of our men have grown rich in politics. I have myself. I've made a big fortune out of the game, and I'm gettin' richer every day, but I've not gone in for dishonest graft—blackmailin'

gamblers, saloon-keepers, disorderly people, etc.

"There's honest graft, and I'm an example of how it works. I might sum up the whole thing by sayin': 'I seen my opportunities and I took 'em.'

"Just let me explain by examples. My party's in power in the city, and it's goin' to undertake a lot of public improvements. Well, I'm tipped off, say, that they're going to lay out a new park at a certain place.

I see my opportunity and I take it. I go to that place and I buy up all the land I can in the neighborhood. Then the board of this or that makes its plan public, and there is a rush to get my land, which nobody cared particular for before.

"Ain't it perfectly honest to charge a good price and make a profit on my investment and foresight? Of course it is. Well, that's honest graft."

(Today that kind of "honest" graft will get you in jail. It's not fair for public officials to take advantage of their knowledge of government business to make a profit.)

Constituents are the people a politician represents.

Fraud is deception and swindling.

constituents can't read," said Tweed, "but, damn it, they can see pictures!" When threats didn't work, Boss Tweed offered Thomas Nast half a million dollars to stop drawing his cartoons. Nast kept drawing. Now people were getting angry about Boss Tweed. Most New Yorkers just hadn't known what he'd been doing.

Tweed was arrested and charged with fraud. He had lied, stolen, and cheated. He was sent to jail. William Marcy Tweed died in jail at age 55. So much for that bad guy.

The state legislature finally passed a third Beach transit bill. But by this time Alfred Ely Beach was a tired man. The stock market was in trouble. It was hard to raise money. Beach no longer had the energy, or the money, to build his grand subway. The subway under Broadway was closed and sealed up.

Beach concentrated on publishing and helping others. Inventors loved him. One day Thomas Edison brought a talking box to him. Beach turned a handle on the box. *Good morning, sir,* said the machine. *How are you? How do you like my talking box?*

Beach spent what money he had left on others. He founded an institute in Savannah, Georgia, to give free schooling to former slaves. He taught himself Spanish and founded a scientific magazine in that language. At age 69 he died quietly of pneumonia, loved and respected by those who knew him.

When the city of New York finally built a subway in 1912, workers tunneling under Broadway were startled to come upon a grand reception room and a small, elegant, wood-paneled subway. Today, scientists say a jet-powered subway in a vacuum tube could whoosh people across the country at amazing speeds. They call it a new idea. Alfred Ely Beach had something like that in mind more than 100 years ago.

"It takes a thief or one who has associated with thieves to catch a thief," said Nast's cartoon. When Nast drew Tweed in prison clothes, the Boss predicted that if people got used to seeing him in stripes they'd end up putting him in jail. They did.

20 Phineas Taylor Barnum

Barnum is supposed to have said about the people who fell for his jokes: "There's a sucker born every minute."

"At the outset of my career," said Barnum, "I saw everything depended on getting people to think, and talk, and become curious and excited over and about the 'rare spectacle.'... Posters...advertisements, newspapers—all calculated to extort attention—were employed, regardless of expense."

A man took a brick from a spot near Phineas T. Barnum's American Museum, walked a block with it, exchanged it for an identical brick, did the same thing with that brick, and another, until finally he carried a brick through the door of Barnum's Museum. Now what's coming may seem strange to you, but it is true. (Remember, this is the 19th century, and there wasn't a whole lot to do on New York's streets.) So people, being naturally curious, followed the strange man to see where he was going to put the brick. Usually a big crowd followed him. When they got to the museum they had to pay admission to get in—and they did. When they got inside the museum they realized they had been fooled: that P. T. Barnum had outwitted them and made them enter his museum with his brick trick. But, instead of being angry, most of them laughed. Barnum called himself "the Prince of Humbug," and a *humbug* is a hoax or an impostor. P. T. Barnum admitted that he became famous fooling people. No one seemed to mind. Barnum was probably the most successful showman America has ever had.

In his American Museum were: a collection of stuffed animals, paintings, a family of trained fleas, musicians, a "real" mermaid, and assorted oddities. It was the biggest tourist attraction in New York. There weren't many museums in those days, and most of Barnum's exhibitions were special, but there was also much tomfoolery. One door was marked TO THE EGRESS. Well, most people had no idea what an egress was so they went through that door thinking they would find a strange animal or who-knows-what. *Egress*—in case you don't know—is another word for *exit*. Once they were through the door they were outside and had to pay to get back in again—and they did. P. T. Barnum had a

Barnum relied heavily on "human curiosities"—above, left, are Chang and Eng, Siamese twins (they both married, and spent three days with one wife and three with the other). At right is Charles Stratton, better known as General Tom Thumb.

Come see the two-horned jigamaree
And the gen-uine mermaid rare!
The elephants in their Sunday pants
And the dangerous polar bear!
This way, this way, for the freaks at play
And the cold pink lemonade!
For Barnum's fooling the world again!
Barnum's on parade!
—STEPHEN VINCENT BENÉT,
A BOOK OF AMERICANS

knack for fooling people and for taking their money. He fit right in with the times.

It was a time when some big business-men—called *robber barons* after the wicked nobles of medieval England who fought each other and plundered everybody else —acted as if they owned the country, and they practically did. It was also a time when some people—called *reformers*—tried to make the country better, and did a pretty good job of that.

But most people in America—after the Civil War—needed to laugh. Phineas Barnum let them laugh at others and at themselves. Since most Americans couldn't get to New York to the American Museum, Barnum took his showmanship and laughter around the country. He built a circus—the Barnum and Bailey Circus. He called it "The Greatest Show on Earth," and, no doubt about it, it was.

The word *circus* comes from the Latin word for "round." (*Circle* comes from the same root.) Romans had circuses. So did the Egyptians, way back, more than 4,000 years ago. But it wasn't until 1770, in London, that there was a modern circus, with a ring and horseback acts and clowns. P. T. Barnum made circuses bigger and better. He introduced three rings (which meant there was no way to get bored) and he built a circus so large that 20,000 people could sit under one big tent. Then he took his circus out to towns across America.

It was lucky for Barnum, and for people throughout the country, that railroads were beginning to cross the land. Barnum used trains to transport his show. It took 60 railroad cars to carry the whole circus. The railroads ran special trains to bring spectators to the circus towns.

America was still mostly farms then, and, of course, there was no radio or television. So you can see that when the circus came it was the biggest event of the year—especially for children. The circus would unload at the railroad tracks and march through town in a big parade. Close your eyes and imagine horses, camels, elephants, lions, tigers—all decorated with ribbons, sparkles, and banners. Then try and see clowns, musicians, and trapeze artists—all in fancy costumes. Picture 300 horses and about 400 performers. Can you understand why boys and girls waited all year for the circus? Hamlin Garland, who was a boy of those days, remembered that

Each year [the circus] came from the east trailing clouds of glorified dust and filling our minds with the color of romance….It brought to our ears the latest band pieces and taught us the popular songs. It furnished us with jokes. It relieved our dullness. It gave us something to talk about.

It also added a word to the English language. Do you ever say something is "jumbo sized" when you mean it is large? The star of Barnum's circus was an elephant whose name was Mumbo Jumbo—but everyone just called him Jumbo. Jumbo was said to be the largest elephant in captivity. Barnum bought him from a London zoo. After the deal was announced the English people were furious. A wealthy American was buying their favorite elephant! Newspapers on both sides of the ocean were full of the story. The English people tried to keep Jumbo on their island. Of course, all that publicity was just fine with P. T. Barnum. He knew Americans would line up to see the elephant that was causing all the fuss. And, of course, they did. Jumbo didn't disappoint them. He was more than 11 feet tall. He also happened to be sweet as a lamb.

Jumbo the Elephant

The circus comes to town. Transporting the show took 60 or 70 freight cars, six passenger cars, and three engines.

The word **jumbo** may have come from the Gulla word *jamba*, which means elephant. Jumbo weighed six and a half tons.

He flies through the air with the greatest of ease,
This daring young man on the flying trapeze,
His figure is handsome, all girls he can please,
And my love he purloined her away!
—GEORGE LEYBOURNE, "THE MAN ON THE FLYING TRAPEZE," 1860

My Regards to Broadway

It was September 12, 1866, and the curtain rose on America's first blockbuster musical show. It was called *The Black Crook,* and one critic said it was "silly and trashy," but the public didn't mind—they loved what they saw. Onstage the scenery was a splendid work of art, and the costumes were more gorgeous (and more expensive) than anything anyone had seen before in the New York theater. Besides, the producer was lucky. A troupe of French dancers had been stranded in New York—they helped make the show "the event of this spectacular age." It ran for 475 performances and was the beginning of the Broadway musical theater; that became America's most original contribution to the world of theater.

Vaudeville performers had to be versatile—able to sing, dance, act, and tell jokes.

The "Mulligan Guards' Plays," which were performed in the 1880s, were the next big Broadway musicals. They were about the different kinds of people found in New York City and their problems and triumphs. (In *The Leather Patch* a henpecked husband escapes his wife by pretending to be dead.) The plays—each had 10 or 12 songs—had handsome sets and costumes and were crowd-pleasers.

In the 20th century, the Broadway musical triumphed with shows like *Kiss Me, Kate, Oklahoma!,* and *Guys and Dolls* by writers and composers with names like Irving Berlin, Cole Porter, Jerome Kern, George Gershwin, Richard Rodgers, Oscar Hammerstein, Alan Jay Lerner, and Stephen Sondheim. See if you can find recordings of their works—you may find that you know some of their songs.

Jumbo captured the heart of the nation. Then he was hit by a train. Grown people wept. Children wept. The tears didn't help: Jumbo was dead. Barnum sent the giant pachyderm's bones to the American Museum of Natural History in New York, and he sent the skin to Tufts University near Boston (you can see still see them in those places today). But he was never able to find another elephant to match Jumbo.

P. T. Barnum got rich and built himself a fantastic home with pagoda-like wings and Gothic turrets. It was as bizarre as you would expect his house to be. But there was a serious side to Barnum. He became a congressman and was a strong supporter of equal rights and of the 14th Amendment (which, as you know, made the laws equal for everyone of every color). And he fought for *prohibition*—which means he tried to outlaw the sale of liquor. His last words were: "How were the receipts today at Madison Square Garden?"

Phineas Taylor Barnum fit the age in which he lived. He combined its extremes: an obsession with money and a desire to do good. But his real gift was his ability to make people laugh—especially at themselves.

It's summertime, and Barnum & Bailey comes to Albany, New York. One friend said of Barnum that he was "the greatest genius that ever conducted an amusement enterprise in this country....never once did I see him falter in anything he set out to do."

21 Huck, Tom, and Friends

Twain and his wife, Olivia. "Grief can take care of itself," Twain said, "but to get the full value of a joy you must have somebody to divide it with."

Some people say that Mark Twain's book *Adventures of Huckleberry Finn* is the best book ever written in America. And they may be right. No question, it is very good reading. So are *The Adventures of Tom Sawyer, The Prince and the Pauper, Pudd'nhead Wilson*, "The Celebrated Jumping Frog of Calaveras County," and his other books and stories.

Twain was like a storybook character himself. When he was an old man and had bushy white hair and a bushy white mustache, he dressed in a rumpled white suit and went around the country telling stories and jokes. People came from far away just to hear him. He had a wry sense of humor and you never quite knew if he was telling the truth or pulling your leg.

"When I was young," he said, "I remembered everything about my life whether it happened or not." Actually, when he was young he had the kind of adventures that boys had in a Mississippi River town, and later he put most of them into his books. He liked to pretend, and maybe that was why he changed his name. Or maybe it was because he ran away from the army during the Civil War and didn't want to get caught. Anyway, his name wasn't Mark Twain at all; his real name was Samuel Langhorne Clemens.

Sam Clemens grew up in Hannibal, Missouri, a town on the banks of the wide Mississippi River. Back then, Hannibal was a sleepy place until a riverboat appeared. Then the town came to life. All the boys in Hannibal wanted to be riverboat pilots when they grew up. Sam became one—for a while. Riverboats sometimes got stuck in shallow water. It

> It's better to keep your mouth shut and appear stupid than to open it and remove all doubt.
>
> —MARK TWAIN

The Clemens girls with their dog, Flash; from left to right, Clara, Jean, and Susy, the oldest, who died tragically of spinal meningitis at age 24.

Mark Twain's writing made him rich, but he never got fancy ideas or forgot where he came from.

Susy and her father clowning as the mythical Greek lovers Hero and Leander. Leander had to swim the Hellespont to kiss his love, so Twain wore a bathing suit and slung a hot-water bottle around his neck.

was important that the pilot know the depth of the water. The call *mark, twain,* meant water two fathoms (12 feet) deep. That was safe water.

In Hannibal, young Sam Clemens was apprenticed to a printer. The printer was his brother. Do you remember about Ben Franklin and his brother, James, who was a printer? Ben was apprenticed to James. Remember that Ben didn't like his older brother bossing him around? Well, it was the same with Sam Clemens. Sam gave his brother a hard time. But he did learn about the printing trade, and he began to write—mostly humorous stories. Then he drifted around the country, taking printing jobs in St. Louis, New York, Philadelphia, and Iowa—and he caught the wandering bug. Later, he traveled to the Middle East and other faraway places, and wrote about what he saw and heard there. But, before that, he tried to be a prospector and find silver in Nevada, where some people were striking it rich. He never did hit a lode of ore, but he did find a subject for his pen—the mining camps—and he turned them into a book called *Roughing It*.

Then he was off for San Francisco, where he got a job as a newspaper reporter; there he wrote a story about a jumping frog, and it helped

Cable, Nineteenth-Century Style

San Francisco is known for its cable cars. They are streetcars that run on a track and cable and go up and down some of the city's very steep hills. In 1873, the first cable car was getting ready to make a test run down Clay Street hill. But when the operator looked down the hill he got scared and jumped off. Inventor Andrew Smith Hallidie grabbed the controls. The ride went smoothly. If you ride a cable car you'll find the view still scary and the ride still smooth.

The track is a deep groove in the street. Through it runs the cable, a very strong electrified wire rope (Andrew Hallidie was a wire-rope manufacturer, and that gave him the idea for the cable car). The driver is called the gripman because he controls a lever underneath the car that grips the cable and powers the car. Often, in the early years, a wire strand would tangle in the grip and the car wouldn't stop. The gripman had to bang his warning gong so the cable cars in front would keep going and avoid a collision. Then the conductor had to jump out of the runaway car, find a phone booth, and tell the powerhouse to turn off the electricity.

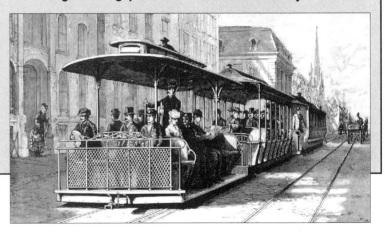

For those used to the clatter of horses on cobbles, the cable car was eerily smooth and quiet.

make him famous. California was a freewheeling place in those just-after-the-war years. Cattle ranching and fruit growing were replacing mining as a source of income. And Nevada's silver, pouring in from the fabulous Comstock lode, was making San Francisco rich. That city was the literary capital of the West; it was a good place for a young writer to be. Clemens became friends with Bret Harte, who wrote stories of the wild side of California—of mining camps and gamblers. Twain said that "Bret Harte trimmed and trained and schooled me patiently."

Have you ever dreamed of being somewhere you know you can't really go? Mark Twain did. He wished he could be in King Arthur's court. He made up a character, put him there, and wrote a book about it. He called the book *A Connecticut Yankee at King Arthur's Court*. (He wrote that book late in his life, when he was living in Hartford, Connecticut.)

Twain knew some American history, and he knew a true story about Christopher Columbus that you probably know: the story about how Columbus was marooned on an island with Indians who wanted to kill him. When Columbus realized that there was soon to be an eclipse of the moon (scientists in Spain had predicted one), he pretended to be a great wizard. He pretended to cause the eclipse.

Twain took that story and let his Connecticut Yankee character live it in King Arthur's court with Merlin the magician and...well, you'll just have to read the book if you want to find out what happened.

He had a way of writing that made people chuckle and then realize that what he was writing was really serious. Mark Twain had something to say. And what he said was that this land of America was pretty terrific but that the promise of America—to offer freedom and opportunity to all—was not being met. There was unfairness in the land, like the unfairness of segregation and child labor. Americans were becoming too

"A pilot, in those days," wrote Twain of the riverboatmen in *Life on the Mississippi*, **"was the only unfettered and entirely independent human being that lived on the earth."**

In *Roughing It*, the story of Twain's western adventures, he got a "bargain horse" which he had to give away after reaching "unexpected elevation."

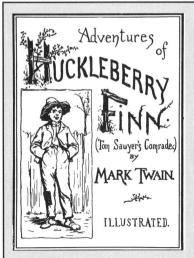

The cover of the first edition of Huckleberry Finn, *published in 1885.*

You don't know me, without you have read a book by the name of *The Adventures of Tom Sawyer*, but that ain't no matter. That book was made by Mr. Mark Twain and he told the truth,

A Great American Book

mainly. There was things which he stretched, but mainly he told the truth." With those sentences, Mark Twain began *Huckleberry Finn* and changed the direction of American literature.

Huck Finn's voice is the voice of a real American boy as it might have been heard in a 19th-century Missouri town. Twain said, "My works are like water. The works of the great masters are like wine. But everyone drinks water."

Another great writer, Ernest Hemingway, said, "All modern American literature comes from one book by Mark Twain called *Huckleberry Finn*. There was nothing before. There has been nothing as good since."

Tom yells at Aunt Polly to look behind her and escapes over the fence.

One of Mark Twain's best books is *Life on the Mississippi*. It is said to be the first book ever written on a typewriter. Twain bought a typewriter from Philo Remington. It had only capital letters and was operated with a foot pedal, but he liked gadgets and it was faster than handwriting. Soon inventors had fancier typewriters for sale, and the American office began to change.

concerned with making money: the nation was forgetting its ideals.

Mark Twain and his writer friend Charles Dudley Warner named the years after the Civil War the "Gilded Age." They didn't mean it as a compliment. Shakespeare wrote a verse that Mark Twain knew. Here it is:

> To gild refined gold, to paint the lily,
> To throw perfume on the violet,
> To smooth the ice, or add another hue
> Unto the rainbow, or with taper-light
> To seek the beauteous eye of heaven to garnish,
> Is wasteful and ridiculous excess.

What did Shakespeare mean by those lines? Read them a few times if you're not sure. Twain believed he lived in a time of "ridiculous excess." He thought many Americans were gilding gold and painting lilies. He was right. In the after-the-Civil War years there was a lot of glitter and gaudiness. People were making money and spending it on show-off things.

When Twain wrote *Huckleberry Finn* he created two heroes: a slave, Jim, and a boy, Huck, who were both searching for freedom. It is a funny book and an adventure story, too, but really it is about the wonder of simple things: of friendship, of a great river, and of the wish to be free.

Mark Twain thought children could teach adults some lessons, so he did a lot of writing about young people. Here is something about Samuel Clemens himself in his own words, from his *Autobiography*:

I was born the 30th of November, 1835, in the almost invisible village of Florida, Monroe County, Missouri....The village contained a hundred people and I increased the population by 1 per cent. It is more than many of the best men in history could have done for a town. It may not be modest in me to refer to this but it is true. There is no record of a person doing as much — not even Shakespeare. But I did it for Florida and it shows that I could have done it for any place — even London, I suppose.

Recently someone in Missouri sent me a picture of the house I was born in. Heretofore I have always stated that it was a palace but I shall be more guarded now.

You can get an idea of Sam Clemens and of Missouri in the mid-19th century from the *Autobiography*. Sam spent his boyhood in the antebellum (before the Civil War) times. After the war people liked to think of those years as uncomplicated and idyllic, which wasn't quite true. If there had been no problems, there would have been no war. But for a boy in a river town it seemed a mighty good time. Here is more of Sam's story:

Mark Twain catapulted to fame in 1865 with the publication of his story "The Celebrated Jumping Frog of Calaveras County."

All of the houses were of logs — all of them, indeed, except three or four; these latter were frame ones. There were none of brick and none of stone. There was a log church, with a puncheon floor and slab benches. A puncheon floor is made of logs whose upper surfaces have been chipped flat with the adz. The cracks between the logs were not filled; there was no carpet; consequently, if you dropped anything smaller than a peach it was likely to go through. The church was perched upon short sections of logs, which elevated it two or three feet from the ground. Hogs slept under there, and whenever the dogs got after them during services the minister had to wait till the disturbance was over. In winter there was always a refreshing breeze through the puncheon floor; in summer there were fleas enough for all.

Sam spent summers at his Uncle John's farm.

Creede, Colorado, one of the mining towns that boomed overnight and that Twain wrote about in *Roughing It*.

His farm has come very handy to me in literature once or twice. In Huck Finn *and in* Tom Sawyer, Detective *I moved it down to Arkansas. It was all of six hundred miles but it was no trouble; it was not a very large farm — five hundred acres, perhaps — but I could have done it if it had been twice as large. And as for the morality of it, I cared nothing for that; I would move a state if the exigencies of literature required it.*

It was a heavenly place for a boy, that farm of my Uncle John's. The house was a double log one, with a spacious floor [roofed in] connecting it with the kitchen. In the summer the table was set in the middle of that shady and breezy floor, and the sumptuous meals — well, it makes me cry to think of them. Fried chicken, roast pig; squirrels, rabbits, pheasants, partridges, prairie-chickens; biscuits, hot batter cakes, hot buckwheat cakes, hot "wheat bread," hot rolls, hot corn pone; fresh corn boiled on the ear, succotash, butter-beans; butter milk, sweet milk, "clabber," watermelons, musk melons, cantaloupes — all fresh from the garden; apple pie, peach pie, pumpkin pie, apple dumplings, peach cobbler.

Exigencies are demands; *sumptuous* means richly abundant.

Nothing so much needs reforming as other people's habits.
—MARK TWAIN

Put all your eggs in one basket and— *watch that basket.*
—MARK TWAIN

That well-fed Missouri boy grew to be a famous man and one of the best writers this country has ever produced. He seemed to have the whole country tucked into the pockets of his white suit. Maybe it was because he'd been everywhere, from Hawaii to Connecticut. Or maybe it was his youthful, questioning, homespun vitality that surprised and delighted people. Even when he was an old man—looking like a white polar bear—he could still think like a child, which isn't a bad thing. He wrote of the world honestly, directly, and with a lot of humor. He didn't ignore bad things, not at all; sometimes he could get downright heavy about them; but, mostly, he made us Americans think about who we are and what we want to be.

Mark Twain loved to play billiards; he did most of his writing in the billiard room. This shot apparently stymied him, and he wrote on the photo: "Oh, John the Baptist couldn't get a count out of an arrangement like that!"

22 Immigrants Speak

Uncle Sam promised immigrants "no oppressive taxes, no expensive kings, no compulsory military service, no dungeons."

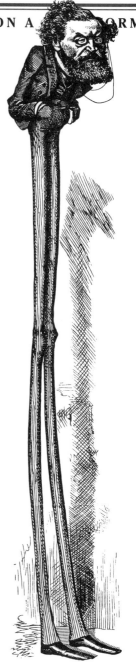

Carl Schurz as drawn by Thomas Nast. In a speech to the Senate in 1872, Schurz said, "Our country, right or wrong. When right, to be kept right; when wrong, to be put right."

Back in 1608—not long after the first settlers arrived in Jamestown—John Smith wrote to his bosses in the London Company and told them the kind of settlers to send to America:

When you send again I entreat you rather send but thirty carpenters, husbandmen, gardeners, fishermen, blacksmiths, masons, and diggers up of trees, roots, well provided; than a thousand of such as we have: for except we be able both to lodge them, and feed them, the most will consume with want of necessaries before they can be made good for anything.

During the 19th century, the kind of people John Smith wanted—working people who could build and farm and invent—came to America in ship after ship after ship. And just in time; with all the new technology there was plenty for them to do. They were needed to work in the new factories, to settle land, and to invent things. Immigrants did all that.

Germans made up the largest single group of 19th-century immigrants. Carl Schurz was one of them, and he went, with remarkable speed, from immigrant to national leader. Here are Schurz's own words telling his story:

It is one of the earliest recollections of my boyhood....One of our neighboring families was moving far away across a great water, and it was said that they would never again return. And I saw silent tears trickling down weather-beaten cheeks, and the hands of rough peasants firmly pressing each other, and some of the men and women hardly

Sometimes the men of the family came over first and sent for the women later. "The great majority are young men and young women, between 17 and 30," wrote one observer in New York, "good, youthful, hopeful, peasant stock."

It is an old dodge of the advocates of despotism throughout the world, that the people who are not experienced in self-government are not fit for the exercise of self-government…[but] liberty is the best school for liberty, and self-government cannot be learned but by practicing it. This, sir, is a truly American idea; this is true Americanism, and to this I pay the tribute of my devotion. —CARL SCHURZ

able to speak when they nodded to one another a last farewell. At last the train started into motion, they gave three cheers for America, and then in the first gray dawn of the morning I saw them wending their way over the hill until they disappeared in the shadow of the forest. And I heard many a man say, how happy he would be if he could go with them to that great and free country, where a man could be himself.

Carl continued:

That was the first time that I heard of America, and my childish imagination took possession of a land covered partly with majestic trees, partly with flowery prairies, immeasurable to the eye, and intersected with large rivers and broad lakes—a land where everybody could do what he thought best, and where nobody need be poor, because everybody was free.

Schurz fought in a freedom movement in Germany (in 1848), but, when the freedom fighters lost, he was in trouble and had to flee to Switzerland. Then, being uncommonly brave, he went back into Germany to help his college profes-

112

sor escape from jail. But he knew if he stayed in Germany, he too would be jailed. Schurz was 23, and he set out for the land of freedom.

When Schurz arrived in America, before the Civil War, he found that some people weren't free. He wasn't the kind of person who kept quiet about something he thought was wrong. After all, he'd been a freedom fighter. He hated slavery, and he spoke out and said so. He reminded people that there was no freedom of speech in the slave states. Without free speech, said Schurz, no one is free, neither slave nor master.

"I am an anti-slavery man, and I have a right to my opinion in South Carolina just as well as in Massachusetts....If you want to be free, there is but one way," said Schurz. "It is to guarantee an equally full measure of liberty to all your neighbors."

In his new country Carl Schurz found the opportunity he had dreamed of in Europe. He studied law, moved to Missouri, and became active in politics. Just 10 years after he arrived in the United States, President Abraham Lincoln named him American minister to Spain. But he soon came home to serve as a general in the Union army. He became a newspaper correspondent, an editor, a U.S. senator from Missouri, and secretary of the interior. He was a public official who talked of conservation of the wilderness and fairness to Indians when hardly anyone else thought of those things.

Like so many other immigrants, Carl Schurz had fallen in love with the ideals of the Declaration of Independence and the guarantees of the Constitution.

Next to the Germans, the Irish were the largest immigrant group. Before the Civil War, one fourth of the whole population of Ireland came to America. (That was 1.7 million Irish men, women, and children.) They kept coming, during and after the war. The Irish were desperate because in Ireland crops had failed, especially the potato crop. There was a famine. More than one million people died of starvation in Ireland. The Irish had

Mathilde Franziska Anneke

Mathilde Franziska Anneke published a liberal newspaper in Germany. The newspaper criticized the government. It was 1848, and some Germans tried to lead a freedom revolution, as Samuel Adams and Patrick Henry had done in America. Mathilde and her husband were among them. But the revolution was squashed. Like Carl Schurz, the Annekes had to flee Germany. They made it to the United States—and freedom. In New York, Mathilde Anneke began publishing a woman's journal. Then she moved to Milwaukee, Wisconsin, where she taught at a school for women and worked for women's rights.

Street life on New York's Lower East Side was noisy, dirty, sometimes dangerous. But it was better to be outside than stuck in a dark, filthy, airless tenement.

113

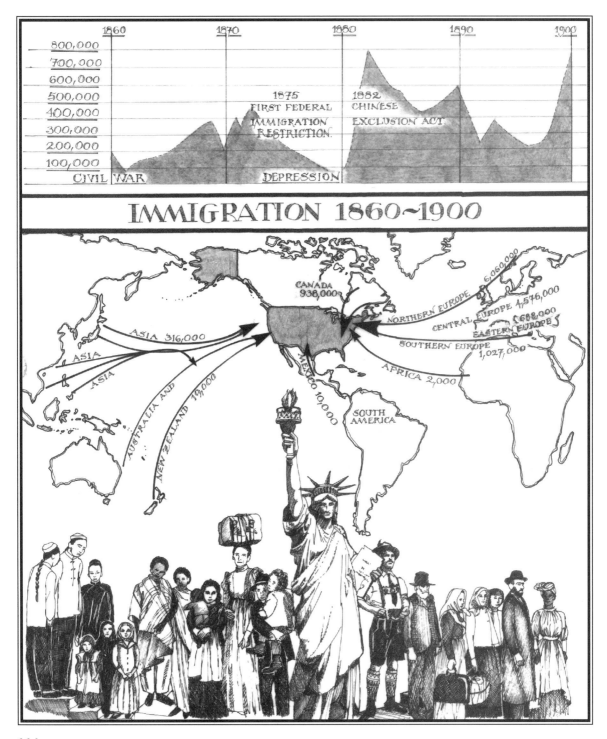

IMMIGRATION 1860~1900

an advantage when they reached America: they could speak the language. So could the Scots, and those from England.

Most of the other immigrants could not. But America's free public schools, which developed in the 19th century, soon taught their children to speak the language. People were now coming from countries that had not sent many people to America before: countries such as Russia, Italy, Poland, Denmark, Sweden, and Hungary.

The population of Europe doubled between 1750 and 1850. All those extra people needed food, homes, and jobs— and there just didn't seem to be enough of them in Europe. Many Europeans came to America to find work and to avoid hunger.

Others came for religious freedom. Religious dissenters came from Holland and Jews (who were often persecuted for their beliefs) came from Germany, Poland, and Russia. Still others came to escape political wars that were leaving parts of Europe in turmoil.

In the half century after the Civil War, some 26 million immigrants arrived in the United States. Think about that—26 million is a whole lot of people. Many of the newcomers began life in the cities—in overcrowded apartment buildings called *tenements*—in conditions that would be illegal today. Sometimes eight families shared one bathroom. Imagine that you are one of them. (Maybe some of your ancestors were.) You don't know the English language, and everything in this land seems different and strange. At school almost everyone speaks English. How do you like it here?

Most of the immigrants knew very little about America except that it was a land of freedom. But that's what they wanted: freedom and a chance to work.

They came on steamships. If they were poor, they were crowded into below-deck areas called the *hold* or *steerage*. It wasn't pleasant down there, but the trip cost $30 and took only 10 days, instead of a month or more as it had in the old-fashioned sailing ships. Most of the immigrants came into New York harbor, to a place called Castle Garden, and after 1892 to Ellis Island (where the early Dutch settlers had picnicked). There they were checked before they could enter the United States. If they had a disease, or if the

A Land of Naked Indians

In Yugoslavia, Michael Pupin sold his warm sheepskin coat to raise money to go to the New World. "Why should anyone going to New York bother about warm clothes?" he said. "Was not New York much farther south than Pancevo [PAN-chuh-vuh], where I had been raised? And when one thinks of the pictures of naked Indians so often seen, does not America suggest a hot climate? These thoughts consoled me when I parted with my sheepskin coat."

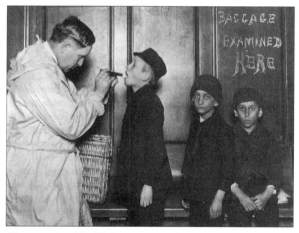

"Are you healthy?" the doctor asked. "Do you have any rashes? Have you ever had tuberculosis?" If he suspected anything, he scrawled a letter in chalk on your jacket.

Abysmal (uh-BIZ-mul) means "awful." The word comes from *abyss*, which is a bottomless pit. Something abysmal is so bad there seems to be no end in sight, no bottom to the problem, no way out. *Abyssal* is a related word. It means: of, or pertaining to, the great depths of the ocean.

Jacob Riis arrived at Castle Garden in New York in 1870.

We don't know exactly why Europe's population increased so fast in the 18th and 19th centuries. Some reasons were the growth of industry and cities, and improvements in public health and nutrition, but they don't account completely for such a big jump. Historians are still trying to figure this out.

papers they brought from their former country were not right, they might be sent back where they came from.

So they were frightened when they arrived.

Bianca de Carli sailed from Italy. Conditions for poor people in Italy were abysmal (especially on the island of Sicily). Some Italians were forced to live in straw shacks and even caves. Many were starving. No wonder America seemed appealing. Still, it took courage to emigrate. Bianca wrote about how she felt when she arrived at Ellis Island:

> A thousand times during the last day or two I put my hands on my passport and papers which I kept wrapped in a handkerchief under the front of my dress. This was just to make sure they were still there.
>
> One of my companions said, "Signora, you are very foolish! When you keep your hand inside your dress…you are telling everyone that your papers and money are there! Maybe a bad person will see. Take your hands away."
>
> Now, years later, I know it was foolish and silly, but we heard so many stories about others who were turned back because their papers were not in order….No one trusted their pockets even, because… crowded together most of the time it would be easy to have our pockets picked.
>
> One woman had sewed her papers and money into the folds of her seventeen skirts! Yes, seventeen; I know I am right in remembering, and she wore them all. She came from a Hungarian province [where] …she told me that a woman's wealth was proved by the number of skirts she could wear.

Jacob Riis was a boy who sailed from Denmark. He had read books about America, and he thought he knew something about the country. What he had read in Denmark were cowboy books. He expected to find buffaloes and cowboys in New York. The first thing he did when he arrived was to take half his

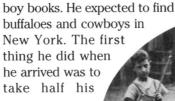

Cities did not yet build public playgrounds. Till then, children had the street—and everything else in it.

Hurrah for America, My Free Country

When 18-year-old Léon Charles Fouquet (foo-kay) headed for America he kept a journal of his adventures. He never did say why he left his home in France, but it is a good guess that if he stayed he would have been drafted into the army. His mother urged him to go to Kansas, where he had some relatives. But first he had to get across the ocean—in steerage, which means the cheapest passage—and the trip was awful. Here are his words to tell you about it:

When our ship left Liverpool, there were many nationalities aboard, but I was the only Frenchman. When the ship anchored at Queenstown, the Irish seaport, it took on a large number of Irish immigrants. They came to the ship in little boats and the waves had such an influence on the little vessels that all those Irishmen got seasick.... *[The seasick Irish came crowding into the dormlike room where Léon slept.]* So offended were my senses that I realized I must rush up on deck to the open air. I had to wait as the little narrow stairway was crowded....At last the stairs were empty and I started up. At the same time a sick man started down. The poor fellow could not control his stomach and out and down it came... right down on me....*[It was an awful start for the voyage and things didn't get much better for Léon.]* The food on the ship was very poor. The bulk of it was like hardtack, hard as rocks, but not too hard for the worms. They were alive in that hardtack and I could not eat. I felt sick. We poor immigrants were treated shabbily.... *[At last the 16-day voyage was over.]* Land! Land! Hurrah! Hurrah for America, my free country! I was jubilant. Everyone on board was jubilant. Oh, how relieved I was as our ship, the *Tariffa*, entered the port of New York during the night on 15 June 1868.

Léon's mother and aunt had told him exciting stories about cowboys and Indians and the "great wonders of the West," and his Uncle Gaillard had sent him $100 to come to Leavenworth, Kansas, and join him. So he bought a train ticket in New York and headed for St. Louis. But when he got to St. Louis the conductor asked him questions in English, and Léon could answer only in French. Things got tense until a man came who spoke perfect French. He was a former slave, from New Orleans, on his way west with his family. Léon was happy to find a friend. But he wanted to see Indians and to ride a horse—and soon he did those things. And, finally, at his uncle's house there was good food: In the center of the table was a golden roasted possum. He thought it a very fine meal indeed. It was the beginning of his life in this free country. Léon married, raised seven children, and became a ferryman, a buffalo hunter, a homesteader, a postmaster, and a justice of the peace. He became an American.

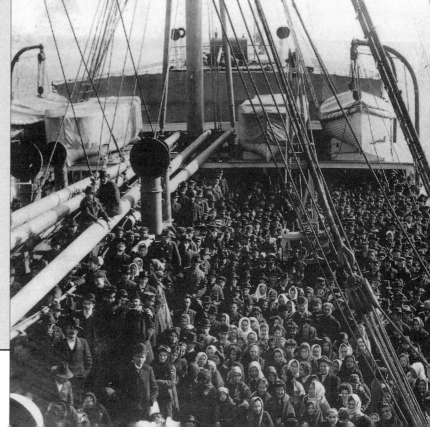

Ships were overcrowded. It was almost impossible to wash yourself, let alone your clothes. Everyone was very glad to see America at last.

money and buy a gun. Later he said he was surprised to find New York "paved, and lighted with electric lights, and quite as civilized as Copenhagen." A friendly policeman saw his pistol tucked in his belt and advised him to leave it home. "I took his advice and put the revolver away, secretly relieved to get rid of it. It was quite heavy to carry around." Riis was very poor for his first seven years in America. Then he got a job as a newspaper reporter and wrote about the difficulties of life for the poor in America's cities. He learned photography. Most photographers took pictures of beautiful scenery or prosperous people. No one was taking pictures of the poor. Riis did. He showed exactly how some people had to live. His books helped get laws passed that made things better. Jacob Riis and Carl Schurz were reformers (and they knew each other). Some Americans, who had been in this country for a long time, had forgotten the nation's founding ideals. But the immigrants had come here to find freedom and opportunity in a land that said *all men are created equal*. They cherished America's ideals.

"Equality of rights…is the great moral element of true democracy," wrote Carl Schurz, who understood exactly what that means. Do you?

Like Bianca de Carli, this family came from Italy to seek a better life. They stayed in New York and opened a grocery in Little Italy. If you visit New York, you will see that much has changed in the last century—but there are parts of Little Italy that still look a lot like this.

23 More About Immigrants

"How could it be," an immigrant recalled, "that after having been so impatient to get there, I suddenly seemed almost frightened…now that we had arrived?"

The immigrants knew the Declaration of Independence well. It was part of what inspired them to come to America. Most of them knew these words by heart: *We hold these truths to be self-evident: that all men are created equal, that they are endowed by their Creator with certain unalienable rights, that among these are Life, Liberty and the pursuit of Happiness.*

Many Americans whose families had been here longer knew those words, too, and believed them and lived by them. But, people being people, there were some who were selfish, or confused, or ignorant, and they ignored that message of fairness.

Some Americans didn't want newcomers in the country; and—this may surprise you—some of the newcomers, as soon as they got settled, didn't want any other immigrants to come. Usually the newest immigrants were poor, and willing to work hard and for less money than those who had arrived earlier. So some people wanted to stop immigration because they feared competition for jobs.

There were other reasons. Because the newcomers were poor and couldn't speak the language, they needed extra help in school. That cost money—tax money. The cities where many newcomers lived were overcrowded and filled with crime, so there was a need for extra police and extra city services. That cost money—tax money. Some people said, "Why should we have to pay for the problems of those poor people?"

IMPORTED, DUTY FREE, by TRUST, MONOPOLY & CO. TO COMPETE WITH AMERICAN LABOR.

A cartoon that reflected the attitude of many workers, who saw new immigrants as cheap competition for their labor.

Eleven excited, newly arrived Dutch immigrants—and their exhausted mother.

The Ku Klux Klan began in Pulaski, Tennessee, as a social club for Civil War veterans. It soon changed its focus. The KKK became dedicated to the idea of white supremacy. White-robed Klansmen, riding out at night, used terror tactics to intimidate blacks and whites who believed in Reconstruction.

They didn't stop to think that the newcomers were often doing jobs no one else wanted to do—washing dishes, or digging ditches, or building railroads. They could not foresee that the sons and daughters of poor immigrants would become some of the most productive citizens any country has ever known. This nation of ours was still young and had a lot of growing up to do. And growing up means making mistakes and learning. Americans made some big mistakes.

There was one mistake that was hateful, hurtful, pernicious, and obnoxious (you get the idea—it was awful). It was the mistake of prejudice. Some Americans faced discrimination—sometimes vicious discrimination—because they were Catholic, Jewish, black, Irish, or Asian.

One group of prejudiced people actually formed a political party. Officially it was named the American Party, but most people called it the "Know-Nothing Party." (What do you think of that name?) The Know-Nothings were anti-Catholic and anti-foreign. They even managed to destroy a stone block sent from Rome by the Catholic Pope. It was intended as part of the Washington Monument.

Another group of haters, the Ku Klux Klan, was anti-black and anti-Semitic (which means they hated Jews). On the West Coast, the Workingmen's Party had as its slogan THE CHINESE MUST GO. Its members hated Asians.

Between 1849 (when gold was discovered in California) and 1882, about 300,000 Chinese emigrated to America. In 1882 there were just over 50 million people in the United States, so 300,000 was a small percentage of the total (what percentage?), but that didn't matter to the haters.

Like many other immigrants, the

Out West, fire fighting was important and teams held competitions. In 1888 the winning team in Deadwood, S.D., was Chinatown's.

Chinese were coming to the United States to make money. American workers wanted high wages to work in the fields or build railroads. Some employers discovered that Chinese men were willing to work for one dollar a day. Now that doesn't sound like much, and it wasn't, but in China times were hard. You had to work several days to earn as much as a dollar—when you could find a job. If a Chinese man came to America and saved $500, he could go home and be prosperous. If he saved $1,000 he went home very rich. Some Chinese did just that. But many found life difficult in America. So the Chinese, too, began asking for more than one dollar a day.

Then some employers sent boats to Japan and got Japanese workers to come and work for low wages. When the Japanese started asking for more money, the employers sent boats to India.

Some said that the employers were exploiting the workers—taking advantage of them. But the process worked both ways. When the Asians took their earnings and went home, they were exploiting the land of opportunity. Many immigrants to the East Coast did that, too. They came, worked hard, saved money, and went back to Greece, or Italy, or Poland— where their American earnings made them seem rich.

It was all right. America had opportunities to share. Besides, most took more than money back to their old worlds. They took American ideas with them too. Ideas of individual dignity, of liberty, and of a government founded to help people pursue happiness. Almost everyone understood those ideals—and they also understood that, although there were haters and bad apples around, most people were good. So people from China kept coming to the land they called the Golden Mountain—as long as they could.

The Chinese come from an ancient, proud civilization.

Advertisements could be racist, too—this one urged patriotic Americans to get along without Chinese labor and buy Magic Washer soap.

Talk about going to the land of the
Flowery Flag made my face fill with
happiness.
With hard work pieces of gold
were gathered together.
Words of farewell were said to the
parents, and my throat choked up.
Parting from the wife, many tears
flowed face to face.

—A VOYAGER NAMED XU
(WHO WENT FROM XIANGSHAN TO SAN FRANCISCO)

Japanese immigrants arrive in San Francisco.

A ***depression*** is a pit or sunken area. "Depression" also describes an emotional state of deep sadness and melancholy. But neither of those meanings has much to do with U.S. history. The *economic* meaning of depression does. A depression is a time when the nation's economy goes into drastic decline and there is much unemployment, low prices, and not much business activity.

When Columbus was a boy, Chinese goods and inventions were the wonder of the world. A Chinese philosopher named Confucius taught ideas about honesty, fairness, and loyalty that were similar to the ideas that most Americans believed in. He also taught about the need for balance in life: between work and pleasure, between consideration of oneself and consideration of others. The Chinese respect learning, and have strong family ties.

Now you might think they would be welcomed when they arrived in California, especially as they came seeking opportunity (which meant jobs). It was the same reason that brought others to California.

But in the 1860s and '70s, when times were difficult in China, it happened that they were difficult in the United States, too. There was a depression, and during depressions there usually aren't enough jobs to go around. Since the Chinese were willing to work hard for very low wages, they usually found work. That angered many white workingmen. Mobs attacked and killed Chinese people; hoodlums burned Chinese homes and laundries.

The Workingmen's Party demanded a law to end Chinese immigration. Congressmen in the East, needing political support from Californians, helped pass that law. Most Americans on the East Coast didn't know any Chinese. They had heard terrible—and untrue—stories of the Asians; many people believed the stories because they didn't know any better.

Most Chinese men braided their hair into a long pigtail. That seemed strange to people who weren't Chinese. Prejudice against those who look different from you is *racism*. Racism is found in almost every nation in the world, and it always leads to evil action. In 1882, American racists got a Chinese Exclusion Act passed. It stopped most Chinese immigration into the United States. It was an especially unfair act considering that it came after Chinese had toiled and died to build railroads and dig mines and labor on farms. Asian immigration was restricted until the 1950s.

"Welcome to the United States, the temple of liberty," said this cartoon—"unless you are Chinese."

24 The Strange Case of the Chinese Laundry

Chinese laundries sprang up everywhere; this one was in Idaho. Most buildings housing them were of wood, the cheapest material available.

Sheriff Hopkins entered the Yick Wo laundry in San Francisco. The sheriff had a warrant for the arrest of the owner. The warrant said Yick Wo had broken the law.

A San Francisco ordinance (*ordinance* is another word for a law, usually a local law) said that all laundries must be placed in brick buildings. The Yick Wo laundry was in a wooden building. The law seemed to make sense. San Francisco had grown quickly after gold was discovered at Sutter's Mill in 1848. Houses and stores were crowded together, and most of them were built of wood. A fire could spread rapidly in that setting.

Laundries used fires to heat water. But then fire was used to heat and cook in almost every building in the city. The fire wardens had inspected the Yick Wo Laundry. It was certified: "In good condition."

San Francisco had 320 laundries; 310 of them were in buildings of wood. Most of the laundries—240 of them—were owned by citizens of China.

Why were there so many laundries? Why did the Chinese own most of them? Why were the Chinese citizens of China and not of the United States? Here are some answers.

The first had to do with sexism. California had a large male population. It was mostly men who had come to California to dig gold. Traditionally, in Europe and America, women were expected to wash clothes. Most American men would not wash their own clothes!

Chinese men were willing to wash clothes. They soon found that only

In 1876 the *New York World* said the western states were "degenerating into Chinese colonies." Yet San Francisco's Chinatown, these children's home, occupied a mere eight square blocks.

123

Children born in the United States are U.S. citizens, whether their parents are Chinese or Turkish or Russian or Pakistani or Colombian or Nigerian or anything else.

The streets in and around San Francisco's Chinatown were narrow, and most of the tenement buildings were wooden.

A person born in another country—and who then comes here, takes out citizenship papers, and becomes a citizen—is a naturalized citizen.

a few people got rich in the goldfields. But miners needed to eat, and they needed to have their clothes washed. It took very little money to open a restaurant or laundry. Many Chinese did those things.

Why didn't the Chinese become citizens of the United States? They were not allowed to become citizens. A law passed in 1790 said that only white people could become naturalized citizens. The law had been aimed at African-born persons. As it turned out, that old law was applied only to Asians. (According to the Constitution, anyone born in the United States is a citizen—no matter where he or she comes from. This law was about naturalization for the foreign-born.)

But that wasn't enough for the racists. They got mean-spirited laws passed to try and put the Chinese launderers out of business. Remember, of the 320 laundries in San Francisco, 310 were in wooden buildings. Sheriff Hopkins arrested almost all the Chinese owners of laundries. He arrested only one of the white laundry owners. (That laundry was owned by a woman. What kind of prejudice was that?!) Seventy-nine white men who ran laundries were not bothered by the sheriff.

The Chinese laundry owners went on trial and were convicted and fined. If they didn't pay the fines they were sent to jail. Their businesses were closed. Was there anything they could do? Remember, they weren't citizens. They decided they would try and do something. In the next chapter you will read all about it.

But first, pretend you are living in the 19th century. If you want to understand history, you need to take yourself into past times. Now, go to China and try to become a Chinese citizen. Well, you'll find that you can't do it—unless you are Chinese. You can't become a citizen of Japan—unless you are Japanese. You can't even get working papers in

A scary scene in Denver, Colorado, in 1880. The Chinese were willing to work for low wages. One employer said, "I find this difference: the Chinaman will stay and work, but the white man, as soon as he gets a few dollars, will leave."

those countries. The same thing is true in most European and African nations. In fact, in the 19th century, there is hardly a country that will allow you citizenship except the land of your ancestors' origin.

Even today, many countries do not let people of other ethnic backgrounds become citizens. In the 19th and 20th centuries Americans struggled with the idea of citizenship. What makes a citizen? Were women citizens? The freedom and openness of the United States was unusual. It still is. Most 19th-century people thought that made our country special—but some people didn't like the idea at all.

Exclusive Rights

President Hayes was against the Chinese Exclusion bill and vetoed it. But in the end it was passed anyway.

What was going on in America? Some people were saying this was a nation where everyone was welcome and would be treated equally, and others were trying to put Chinese laundrymen out of business. Have you ever had two ideas at the same time—one good and one not so good? Well, nations are no different from people. That idea that *all men are created equal* was fighting it out with an idea that some people called *nativism* and others called *racism*. Actually, racism has a nasty history in nations all around the world. In the 1880s, nativists in California directed their energy against the Chinese. They spread vicious rumors about Chinese people. They burned Chinese homes and businesses and attacked Chinese people. They persuaded Congress to pass the Chinese Exclusion Act (in 1882). It stopped Chinese people from emigrating to the United States. Then, as if that wasn't enough, laws were passed—like the laundry act—to try to take jobs away from Chinese workers.

25 Going to Court

THE
NEW DECLARATION OF "INDEPENDENCE."
"...FOR TWENTY YEARS NO MORE
CHINESE LABORERS SHALL COME TO THE
UNITED STATES;...AND NO COURT
SHALL ADMIT CHINESE TO
CITIZENSHIP."

The cartoon caption says: *Fritz [a German] to Pat [an Irishman]*: "If the Yankee Congress can keep [the Chinese] out, what is to hinder them from…keeping us out too?" A good question.

In a criminal case, the prosecution may not appeal if it loses. No person may be tried twice for the same crime. That is called *double jeopardy*, and we don't have to worry about it in this country.

If you decide to be a lawyer, and you go to law school, you will probably study the case of *Yick Wo* v. *Hopkins.*

Here is something the lawbooks won't tell you. Even the Supreme Court didn't know it. There never was a Yick Wo. Sheriff Hopkins made a mistake. He assumed that the man who owned the Yick Wo Laundry was named Yick Wo. Actually, his name was Lee Yick.

We don't know much about Lee Yick, except that he came to California in 1861 and operated a laundry for 22 years. We also know that he was willing to fight for his rights. But what were his rights? He wasn't a citizen. Did he have the same rights as if he had been an American citizen? No one was sure.

After he was arrested, in 1886, Lee Yick went to the other Chinese launderers. He persuaded them to help him hire a good lawyer. If he could get the verdict in his case changed, it would affect them all. They *appealed* his case. Do you know what that means?

Here is an explanation of what an appeal is and a basic idea of how our judicial system works. It may come in handy; most people go to court at some time in their lives.

To begin, there are two kinds of law cases: *civil* and *criminal.* Suppose you buy an expensive bicycle, take it home, and it isn't what you expected. You feel cheated. The bike dealer doesn't agree. He thinks you got just what you paid for. He won't take the bike back. If you can't settle the argument, you can go to court. You can sue the bike seller. It will be a *civil* case. No criminal laws have been broken. It will probably be a judge who decides who is right—you or the bike dealer.

However, in most cases, civil or criminal, you have a right to a *jury trial*. A *jury* is a group of citizens—ordinary people—who listen to the evidence and decide what happened. They decide if someone is wrong, or guilty of breaking a law. Then, in civil cases, the jury decides on the penalty. In criminal cases, if a person is convicted, the judge usually decides on the proper penalty.

The *Yick Wo* case was a criminal case; it was about a San Francisco law. The case began in a local—San Francisco—court. (We have local, state, and federal courts.) Lee Yick was the *defendant*—the person on trial. The other side—the San Francisco authorities—were prosecuting the case. They were the *prosecutors*.

The judge and jury heard arguments by lawyers from both sides. They called *witnesses*: people who had information about the case. Lee Yick's lawyer called the safety inspectors. But the law was clear—and, as you know, the San Francisco jury said that the Chinese launderers

The jury decides on the facts of the case; the judge decides what law applies to the case.

Some people thought that a comet was a sign that the world was coming to an end. And some scared racists thought that Chinese labor (the comet) would bring their world to an end. This cartoon suggests that the press (the giant telescopes) makes things worse by exaggerating the size of the threat.

had broken the law. Lee Yick—alias Yick Wo—had to pay a fine or go to jail.

After a decision is made, a case may be *appealed* to a higher court. Lee Yick's lawyer appealed. The case went to the California Supreme Court.

Appeals courts do not hear witnesses. They do not have juries. They aren't like the courts you see on TV. The job of the appeals court is to review the lawyers' arguments and see if the law has been applied properly in the lower court. The court asks: *Has justice been served?*

The California Supreme Court agreed with the lower court. It said the decision in *Yick Wo* v. *Hopkins* was correct. The Chinese laundries were in wooden buildings, and wooden laundries were illegal.

You can imagine how Lee Yick and his friends felt. They still thought the city of San Francisco was being unfair. White laundry owners were running laundries in wooden buildings. Why was the city picking on Chinese laundrymen? Lee Yick decided to take his case to the highest appeals court of all—the United States Supreme Court.

The Supreme Court doesn't listen to all the cases that people want it to hear. It couldn't possibly do that. It selects cases carefully. It tries to pick cases that will test important issues, especially constitutional issues.

There were two issues in this case. The first was this: Do the police have the right to enforce a law *arbitrarily*? (ar-bih-TRARE-ih-ly—it means inconsistently; not the same way to everybody, every time.)

The second issue had to do with the rights of noncitizens. Should the law treat *aliens* (people who aren't citizens) the same way that it treats American citizens?

This was an important case. Police departments in many states were interested. They

Artist Thomas Nast shows the Republicans (the elephant) and the Democrats (the tiger) both picking on the Chinese—and uprooting the tree of liberty. (Nast soon gave the Democrats a new symbol—a donkey.)

didn't want their power limited. They wanted the power to treat aliens (those noncitizens) as they wished. *Briefs*—which are written legal arguments—were presented to the Supreme Court by Nebraska, Iowa, Indiana, Mississippi, New Jersey, Wisconsin, and Florida. They were all in support of Sheriff Hopkins. Can you guess what happened?

The sheriff and the states lost the argument. The Chinese laundrymen beat them. This is what the Supreme Court said:

> *For no legitimate reason this body by its action has declared that it is lawful for 80-odd persons who are not subjects of China to wash clothes for hire in [wood] frame buildings, but unlawful for all subjects of China to do the same thing.*

It was a law applied "with an evil eye and an unequal hand." That, said the justices, was wrong. And, said the court:

> *The 14th Amendment to the Constitution is not confined to the protection of citizens. It says: "Nor shall any state deprive any person of life, liberty, or property, without due process of law; nor deny to any person within its jurisdiction the equal protection of the laws."*

Within its jurisdiction: that means that all persons in the United States, citizens or not, are entitled to the same fair treatment. Lee Yick and his friends had won a momentous victory.

In 1981 a San Francisco public school was named Yick Wo Elementary School.

Despite the prejudice they encountered, many Chinese became attached to frontier life in the West and felt at home. Ah Can (left), photographed in the 1920s, was an herbalist and the last Chinese to live in Warrens, Idaho.

26 Tea in Wyoming

If you were a man, would you have argued with Esther Morris? She was six feet tall, weighed 200 pounds—and had a cause she believed in.

Everyone knows about the Boston Tea Party, but have you heard of the Wyoming Tea Party? Well, in Wyoming the tea actually got served, along with cakes and cookies—and all was delicious. The day was pleasant, and the conversation lively, when Esther Morris got to the point of the tea party.

Esther Morris always got to the point. Everyone expected it of her. She was gracious, very intelligent, and dependable—her two guests thought highly of her; so did most people in South Pass City. And, with its 3,000 inhabitants (mostly miners), South Pass City was the largest town in the Wyoming territory.

Esther Morris grew up in Oswego, New York, became an orphan at age 13, learned to make hats, and earned a living as a milliner (a ladies' hatmaker). Then, when her husband decided to go west in search of gold, she went with him. Esther was so capable that she was asked to be justice of the peace. She took the job and became—so it is said—the world's first woman justice of the peace.

Now, back to the tea party. Esther's guests were the two candidates for the Wyoming legislature—a Democrat and a Republican. (She wasn't going to take any chances; she wanted to cover both parties.) She asked each of them to promise, if he was elected, to introduce a bill into the legislature giving women the vote in Wyoming. This was in 1869; no women, anywhere, had the right to vote.

But, as I said, the tea was delicious and Esther Morris was persuasive. The candidates both agreed. When Colonel William H. Bright, the

Some of the male Democrats changed their minds about female suffrage when a Republican was elected. They blamed their defeat on the women. One of them, legislator W. R. Steele, was mighty upset. This is what he said: "Woman can't engage in politics without losing her virtue. No woman ain't got no right to sit on a jury, nohow, unless she is a man, and every lawyer knows it."

Julia, wife of Colonel Bright (above), also put pressure on her husband to get women the vote.

Democrat, got elected, he kept his word. He introduced the bill and got it through the Wyoming Senate. Suddenly, people got wind of what was happening. The women's suffrage bill still had to pass in the Wyoming House of Representatives, and then the governor had to sign it. Now there was a lot of debate and hoopla; it didn't go easily—but it happened.

On November 9, 1869, the legislature passed *An Act to Grant to the Women of the Wyoming Territory the Right of Suffrage and to Hold Office.* Now there was another worry. Would the governor veto the bill? He was a Republican and all the other officeholders were Democrats. Some thought he would. But Governor John Campbell had attended a women's suffrage convention

Women in Colorado, Idaho, and Utah all had the vote by 1900. In 1910, Wyoming's Mary G. Bellamy became the first woman elected to a state legislature. In 1917, Jeanette Rankin, from Montana, was the first woman elected to the U.S. Congress. And, in 1925, Nellie Taylor Ross became the first female governor when she took office in Cheyenne, Wyoming.

Women voting in Cheyenne, Wyoming. The legislature might have worried more about the innovation, except that in 1870 there were six times as many men as women in the territory. But women now had a foot in the door.

131

We Don't Want the Vote

In 1893 Colorado became the second state to give women the vote. But, like Mrs. Rutherford, many women still feared and hated the idea.

Some women were against female suffrage. Mildred Rutherford, who was president of the Georgia United Daughters of the Confederacy, supported the National Association Opposed to Woman Suffrage. This is what she had to say:

The women who are working for this measure are striking at the principle for which their fathers fought during the Civil War. Woman's suffrage comes from the North and the West and from women who do not believe in states' rights and who wish to see negro women using the ballot. I do not believe the state of Georgia has sunk so low that her good men cannot legislate for women. If this time ever comes then it will be time for women to claim the ballot.

back in Salem, Ohio—he'd heard Susan B. Anthony speak, and she made sense to him. He didn't veto the bill. Wyoming's women could vote!

Just what would happen when women went to the polls? No one knew. A newspaper reporter, writing in the *Laramie Boomerang*, quoted a mythical railroad man who said of women's suffrage, "It's a kind of wild train on a single track, and we've got to keep our eyes peeled or we'll get into the ditch.... Female suffrage changes the management of the whole line....We can't tell when Wyoming Territory may be sidetracked with a lot of female conductors and superintendents and a posse of giddy girls at the brakes." Some eastern newspapers made fun of women and of the West. They drew cartoons of tough, cigar-smoking female cowboys lassoing their males. But when election day came, 70-year-old Louisa Ann Swain put a clean apron over her dress, walked to the polls, and became the first woman to cast a ballot in a public election. A voting official wrote, "No rum was sold, women rode to the polls in carriages furnished by the two parties, and every man was straining himself to be a gentleman because there were votes at stake."

Eighteen years later, when Wyoming was about to become a state, some U.S. congressmen objected to its tradition of female suffrage. A group of women was afraid that this objection might keep Wyoming from becoming a state. They telegraphed the Wyoming representative in the nation's capital: DROP US IF YOU MUST. WE CAN TRUST THE MEN OF WYOMING TO ENFRANCHISE US AFTER OUR TERRITORY BECOMES A STATE. But the men of Wyoming wouldn't go along with that idea. They telegraphed the same representative: WE MAY STAY OUT OF THE UNION A HUNDRED YEARS, BUT WE WILL COME IN WITH OUR WOMEN. And they did.

As far as we know, Louisa Swain was the first woman to vote in the United States.

27 Are You a Citizen If You Can't Vote?

The motto of Susan B. Anthony's newspaper *The Revolution*: "Men their rights and nothing more; women their rights and nothing less."

"The best women I know do not want to vote," said Horace Greeley, who was an important newspaperman and politician.

Oh, my. The woman he was saying this to was Susan B. Anthony, a tall, rawboned Quaker who had spent much of her life trying to get the vote for women.

Best women, indeed! *Foolish women*, thought Susan Anthony.

So she went to see President Ulysses S. Grant, who had an intelligent wife, named Julia. Grant believed that women could be postmasters, and had named 5,000 women to that job. "Isn't that enough?" he asked Susan Anthony.

No, it wasn't enough. What she wanted was "justice, not favors."

The vote for women was a hot issue. Grant was running for reelection. Suffrage for black men came first, he said. Women would have to wait. Besides, many sensible people thought women's suffrage would be the end of the family. A husband might vote Republican and a wife Democratic. Could a marriage survive that kind of thing?

No one knew for sure, but it certainly sounded ominous.

But not to people like Susan B. Anthony, Elizabeth Cady Stanton, Lucy Stone, Sojourner Truth, and others who knew that a woman without a vote was not a full citizen.

Horace Greeley never forgave Elizabeth Cady Stanton for getting his wife to sign a suffrage petition.

These women organized the National Woman Suffrage Association in 1869, after the 15th Amendment had given the vote to black men. That's Susan B. Anthony seated in the center.

Those women had worked hard to abolish slavery. They believed that abolition went hand in hand with women's rights. But many male abolitionists—like Horace Greeley—were ignoring women's rights. The women felt betrayed.

The 15th Amendment said:

> *The right of the citizens of the United States to vote shall not be denied or abridged by the United States or by any state, on account of race, color, or previous condition of servitude.*

Schools for Women

Now that women were being accepted on a more equal level with men in academic programs, they could also participate seriously in physical exercise and athletics.

Nineteenth-century women did things American women had rarely done before. They began working in factories, speaking in public, writing for newspapers, and getting an education. Some were staying in school until they finished high school. A few, who were lucky, got to go to college. Mount Holyoke, a women's college, was founded in 1837. West of the Mississippi, the first college for women was opened by Cherokee Indians in 1851. It was the Cherokee National Female Seminary in Park Hill, Oklahoma. (A male seminary was founded at the same time. In 1907 those seminaries became part of Oklahoma's Northeastern State College.) Vassar College was founded in 1865; Smith and Wellesley colleges 10 years later; Bryn Mawr College 10 years after that. Some 19th-century women attended coeducational schools like Oberlin College and the University of California.

On February 12, 1887, a *Harper's Bazaar* item read: "Senator Stanford's university in California will be open to young women as well as young men, and all the laboratories and libraries will be used by the two sexes in common."

Carry Nation was born in Kentucky. She began fighting for temperance after leaving her first husband, a terrible alcoholic.

Anthony and Stanton, who were the nation's leading women's rights leaders, wanted the new amendment to read "race, color, *sex*, or.... " But even sympathetic congressmen said that the amendment would be difficult enough to get passed as it was.

The 15th Amendment was ratified in 1869 and became law in 1870. Now the question was: Who was a citizen? Were women citizens?

As you know, it wasn't a question at all in the territory of Wyoming. Of course women were citizens! Few people in the West pretended that women were dependent on men.

But in the East it was hard to break old traditions. Women had never voted. There was something else too. Many of the women reformers were working for *temperance* as well as the vote for women. (Temperance advocates wanted to put an end to the drinking of liquor.) Some men who might have encouraged women to vote were afraid that voting women would close their bars. They were probably right.

In 1874 the Women's Christian Temperance Union was founded to battle drunkenness. Women went into men-only saloons. Carry Nation went with a hatchet and did some chopping in those places. Carry, like Sojourner Truth, was a powerful, six-foot-tall woman. Men ran when they saw her coming.

But all that was beside the point to Susan Anthony, who believed, as the colonists had in 1775, in "no taxation without representation."

Women could be taxed, but they couldn't vote.

Women could be arrested, but they couldn't serve on a jury.

In 1871, in Washington, D.C., 72 women tried to vote and were turned away at the polls.

That same year three women in Nyack, New York, and one woman in Detroit, Michigan, did vote. The *New York Times* wrote in an editorial, "No evil results followed."

Susan B. Anthony, in Rochester, New York, did a lot of thinking about the 15th Amendment. It said that all citizens could vote. Anthony visited a friend, lawyer Henry R. Selden. He had been a judge

Carry Nation was not the only ax-wielding saloon smasher. She was arrested 30 times, but many temperance supporters got away with it, especially if their husbands were important citizens. This poster saw them as avenging angels.

A Woman of the Church

In the East, women were rarely preachers. But that was less true in the West. Olympia Brown was born in Michigan. She went to Antioch College in Ohio (where Horace Mann was president), and to Canton Theological School (where she became a Unitarian minister). In 1878, she and her husband and two children moved to Racine, Wisconsin. There she took over a church ("given up as hopeless by several men," said Elizabeth Cady Stanton). Brown soon had "large audiences and earnest members about her." She also worked for temperance and woman suffrage.

Back in 1848 at the Seneca Falls Woman's Rights Convention, Elizabeth Cady Stanton (here photographed much later, as an old lady) wrote: "Resolved, That it is the duty of the women of this country to secure to themselves their sacred right to the elective franchise."

on the New York Supreme Court.

Was she a citizen? Could she vote? she asked. Selden thought the answer to those questions was *yes*.

That was all Susan B. Anthony needed to hear. It was November 1, 1872. She and 15 other women marched to a barbershop in Rochester's Eighth Ward. Three men who sat there were registrars. Their job was to register voters. They were stunned by the women's request. The women wanted to be registered to vote.

The three men didn't quite know what to do. They argued a bit, but finally agreed. They saw no harm in women voting.

The next day the newspaper was full of the story. So were newspapers in other parts of the nation. Mostly the newspapers disapproved. "Lawbreakers" were what the women were called.

But on voting day, November 5, the 16 women were at the polls at 7 A.M. When they voted, that big news went by telegraph from Maine to Florida to California to Washington.

Twenty-three days later, on Thanksgiving Day, a tall deputy marshal knocked on Susan B. Anthony's door. He looked uncomfortable and stammered a bit. "Miss Anthony," he said. "I have come to arrest you." He had a warrant in his pocket. It said she had broken an act of Congress.

That day the other 15 women voters were also arrested. They were brought to court.

Anthony was asked if she had gone "into this matter for the purpose of testing the question."

"Yes, sir; I had resolved for three years to vote," she answered.

She was ordered to appear before a grand jury. The three men who had registered the women were arrested for registering and accepting ballots unlawfully.

The government decided to prosecute Susan B. Anthony alone—she would represent the 16 women. The three men would all be tried. It was January. The trials were set for June. They had six months to prepare.

The judge may not have realized the kind of woman Anthony was (or perhaps he did and that was why he was afraid to let her vote). She was a fighter, and unafraid. She used those six months well. She spoke in all of Rochester's districts. She talked about the Constitution

Law and the Ladies

Myra Colby Bradwell passed the Illinois bar exam. That meant she knew everything a lawyer was required to know. But it was 1869 and the Illinois Supreme Court said she could not practice law for one reason it thought important: she was a woman. Bradwell took her case to the U.S. Supreme Court. That court agreed with Illinois (in 1873), declaring, "the paramount destiny and mission of women are to fulfill the noble and benign offices of wife and mother. This is the law of the Creator." Mary Colby Bradwell was out of luck. So were many other women of her time.

When Belva Lockwood ran for president, these New Jersey men tried to mock her by dressing up in women's clothes. But they ended up looking stupid themselves.

But some women just wouldn't stay benign. (*Benign* means harmless and not threatening.) Belva Ann Lockwood was one of them. She studied law at the National University Law School, but—that woman problem again—she wasn't given her diploma. (It was all right for women to study, but not to work in jobs where they might compete with men.) The National University (it no longer exists) was under the wing of the president of the United States. Belva Ann Lockwood wrote President Ulysses S. Grant a letter. This is part of what she said. "You are,

Belva Lockwood

or you are not the President of the National University Law School. If you are its President I wish to say to you that I have been passed through the curriculum of study of that school, and am entitled to, and demand my Diploma. If you are not its President then I ask you to take your name from its papers."

Whew! How would you respond to that if you were the president? Belva Lockwood was allowed to practice law. (It was 1873, the year Bradwell was turned aside in Illinois.) Three yeas later, when Lockwood was

told she couldn't argue a case before the Supreme Court, she lobbied Congress, got a bill passed, and became the first woman admitted to practice before the highest Court.

Being first wasn't unusual for Belva Lockwood. She was the first woman to ride a bike in Washington, D.C. It was a big three-wheeler, and the newspapers commented—unfavorably. She ran for president as a candidate of the National Equal Rights Party in 1884 and again in 1888. "I do not believe in sex distinction in literature, law, politics, or trade," said Belva L., "or that modesty and virtue are more becoming to women than to men; but wish we had more of it everywhere."

Some Remarkable Women

Clara Barton

CLARA BARTON established the American Red Cross.

ALICE HAMILTON was the first woman professor at Harvard Medical School.

ANNA J. COOPER, born in North Carolina to a former slave, earned a master's degree at Oberlin College and a doctorate at the Sorbonne in France. She became principal of Dunbar High School in Washington, D.C., and made it one of the finest schools in the nation.

OTELIA CROMWELL, another descendant of slaves, earned her Ph.D. at Yale after attending Smith College and Columbia University. She became a teacher and writer.

Helen Keller

CHARLOTTE PERKINS GILMAN wrote books that made women question many of their old ideas.

HELEN KELLER, who was blind and deaf, became a speaker and writer of note. She proved there are no handicaps to a person of determination and intelligence.

and natural rights. She used Thomas Jefferson's word, *unalienable*.

Rights were not something that governments owned and gave out to people, she explained. They belong to each of us. People are born with rights. Governments are formed to protect those rights.

She called them "God-given rights."

She also talked about the "hateful oligarchy of sex." By which she meant the rule of men over women. Half the people were ruled by the other half, she said.

"*We, the people* does not mean *We, the male citizens*," she said. And, she added, it was "downright mockery to talk to women of their enjoyment of the blessings of liberty."

The Rochester judge did not agree. But he was smart enough to know that most people in Rochester were being swayed by Susan Anthony's speeches. So he had the trial moved to a little town 28 miles away. Twenty-eight miles was a good distance in horse-and-buggy days.

It was an important case. Everyone knew that. When the day came, the courtroom was packed. A former president, Millard Fillmore, was one of the spectators. Newspaper reporters sat with their pencils sharpened.

Judge Ward Hunt was a foe of women's rights. He wouldn't let Susan Anthony speak for herself. He judged her "incompetent."

Anthony's friend, lawyer Henry Selden, said, "Every citizen has a right to take part upon equal terms with every other citizen." And, he said, "Political bondage equals slavery."

Judge Hunt turned to the jurors. "Under the 15th Amendment …Miss Anthony was not protected in a right to vote.…therefore I direct that you find a verdict of guilty."

Now that was something no judge has the right to do. Judges can tell a jury about the law. They can't tell juries how to vote.

The clerk of the court said to the jury, "You say you find the defendant guilty, so say you all?"

No juror said a word.

Judge Hunt said, "Gentlemen of the jury [of course, there were no women], you are dismissed."

The judge ruled Susan B. Anthony guilty. Not a juror had spoken. Most were outraged.

Now the issue was no longer the vote for women. It was an issue of a free trial in a free society. This trial had been a joke.

Anthony's lawyer reminded the court of Matthew Lyon, who had been imprisoned and fined for saying bad things about President John Adams. (That was during the time of the Alien and Sedition acts, when free speech had been in trouble. Later, the court apologized, and Matthew Lyon's fine was paid back, with interest, to his heirs.) That didn't matter to the judge. Susan B. Anthony and the three men were fined and sentenced to jail.

Judge Ward Hunt

The *New York Sun* wrote of a "jury of twelve wooden [figures] moved by a string pulled by the hand of a judge." A Utica, New York, paper said Judge Hunt had "outraged the rights of Susan B. Anthony" (even though the paper's editor didn't think that women should vote). Another newspaper said, "The right to a trial by jury includes the right to a free and impartial verdict."

Susan B. Anthony refused to pay her fine, and Judge Hunt, perhaps knowing he had gone too far, never demanded it. She did not go to jail. No appeal was ever heard by a higher court.

The three male registrars spent five days in jail. They didn't mind it a bit. While they were there they ate fancy meals sent by the 16 women they had registered. Then President Ulysses S. Grant pardoned them. At the next election, Rochester's male voters reelected them by a large vote.

But Susan Anthony, and America's women, lost out. If she had won that court case, in 1873, women all over the nation would have been able to vote. The word *citizen* in the 15th Amendment could have been interpreted to mean "men and women citizens."

Judge Hunt decided things otherwise. At the time, many men, and women, too, believed he was right. It would take much work—by Susan B. Anthony and others—to change their thinking. It would take another amendment (the 19th) before women had the rights of full citizens.

There will be no end to it [talk of independence]. New claims will arise; women will demand a vote; lads from twelve to twenty-one will not think their rights enough attended to; and every man who has a farthing [a small coin] will demand an equal voice with any other, in all acts of state.
—JOHN ADAMS IN 1776

Was John Adams right?

28 Mary in the Promised Land

"I was going on a wonderful journey," wrote Mary Antin. "I was going to America."

Autobiographies are usually written by gray-haired men or women. But Mary Antin was not yet 30 when she wrote her life story. She had a tale to tell, she told it well, and many were fascinated to read it.

"I began life in the Middle Ages," she wrote. What she meant was that she was born in the Old World, into a way of life that had not changed since medieval days.

It was in Russia that she was born, in a region called "the Pale of Settlement." "Within this area the Czar commanded me to stay, with my father and mother and friends, and all people like us. We must not be found outside the Pale, because we were Jews."

Her little village—called, in her Yiddish language, a *shtetl*—was Polotzk, and its roads were of dirt and some of its houses had dirt floors too. Its people were pious, and set in their ways, for they lived just as others had done for generations before them.

Slowly, Mary came to understand that there was a fence around Polotzk and the whole Pale. It was an imaginary fence, but it kept her and the other Jews prisoners. It allowed those who were not Jews to make fun of her—or do much worse things—and suffer no penalties.

A **pale** was once a real fence or boundary (you may have seen fences built of tall, thin sticks called *palings*). In some European towns the Jewish quarter was walled or fenced off. Even when it wasn't, everyone knew where the boundary began. The expression *beyond the pale* means "going too far."

The first time Vanka threw mud at me, I ran home and complained to my mother, who brushed off my dress and said, quite resignedly, "How can I help you, my poor child? Vanka is a Gentile. The Gentiles do as they like with us Jews." The next time Vanka abused me, I did not cry, but ran for shelter, saying to myself, "Vanka is a Gentile." But the third time, when Vanka spat on me, I wiped my face and thought nothing at all….The world was made in a certain way and I had to live with it. Not all the Gentiles were like Vanka. Next door to us lived a Gentile family which was very friendly. There was a girl as big as I

who never called me names, and gave me flowers from her father's garden. And there were the Parphens....On our festival days they visited our house and brought us presents, carefully choosing such things as Jewish children might accept; and they liked to have everything explained to them, about the wine and the fruit and the candles, and they even tried to say the appropriate greetings and blessings in Hebrew.

When Mary was very small her family owned a store and was prosperous. And though her house might not seem grand to us in America, it seemed fine in Polotzk. She had clothes as nice as her friends, and better than most.

But her father got sick, and then her mother, and soon everything was gone—the store and the nice house and the clothes. They were poor, very poor, and had hardly enough to eat.

Even if they had been rich, Mary could not do what she wanted to do. She had a good mind, and she loved to read, but the schools in Russia were closed to most Jews and there were no public libraries in the Pale.

Because her parents cared about learning, they paid a *rebbe*—a teacher—to teach their daughters.

> *Long before we had exhausted Reb' Isaiah's learning, my sister and I had to give up our teacher, because the family fortunes began to decline, and luxuries, such as schooling, had to be cut off. Isaiah the Scribe taught us, in all, perhaps two terms, in which time we learned Yiddish and Russian, and a little arithmetic.*

Their brother went to religious school, which was only for boys. In Polotzk, once a girl learned to read her Hebrew prayers she was supposed to be content with sewing and household chores. And Jewish boys who wanted to study something beside religious books had no way to do it.

Mary's father had been that kind of boy: a scholar, but not of religion. Perhaps that was why he wanted to bring his family to America. Perhaps it was because he was a failure in the Old World.

Whatever the reason, he made the journey, got off the immigrant ship, went to Boston, and stayed in that venerable city. But he had no talent for

Back home in the "old country," in the *shul* or synagogue, the men and boys prayed and studied Torah and Talmud (the first five books of the Hebrew Bible and the scholarly writings about it). Women and girls sat separately, upstairs.

Between 1860 and 1890, some 10 million Europeans came to the United States. Many came for religious freedom. Others hoped to find economic opportunity. Some came for the political freedom that democracy promised. And some came for all those reasons.

Inside the Pale (which was an area of eastern Europe then ruled by Russia), many shtetls, smaller than the one above, were poor villages. In towns the Jewish quarter was often a slum. Jews could not hold public office; special laws restricted the business Jews could conduct and taxed them more heavily than Christians.

Mary Antin (left) with her friend Elinore Stewart, a Wyoming homesteader who also wrote about her life in America, very different from Mary's.

success. After three years in Boston, he had to borrow enough money to send steerage tickets to his wife and three children. Mary was 12.

When the tickets arrived in Polotzk, Mrs. Antin and her children became celebrities. They were going to America! Everyone wanted to give them advice, or ask favors. Hayye Dvoshe, the wigmaker, hadn't heard from her son, Moshele. She thought he was in "Balti-moreh." Wasn't it in the neighborhood of Castle Garden, where the immigrant ships landed? Would they find her son and let her know how he was doing?

The last night in Polotzk we slept at my uncle's house....But I did not really sleep. Excitement kept me awake, and my aunt snored hideously. In the morning I was going away from Polotzk, forever and ever. I was going on a wonderful journey. I was going to America. How could I sleep?

Half of Polotzk was at my Uncle's gate in the morning, to conduct us to the railway station, and the other half was already there before we arrived....The last I saw of Polotzk was an agitated mass of people, waving colored handkerchiefs and other frantic bits of calico, madly gesticulating, falling on each other's necks, gone wild altogether. Then the station became invisible, and the shining tracks spun out from sky to sky. I was in the middle of the great, great world, and the longest road was mine.

For Mary it may have been a great, great world, but for her mother it was terrifying. She was alone with four children. None of them had ever seen the ocean, or been beyond the Pale, or heard English spoken. It was a difficult journey on a crowded ship.

And when they arrived in America they had come, as Mary Antin wrote, from the Middle Ages into the modern world. For their first American meal:

My father produced several kinds of food, ready to eat, without any cooking, from little tin cans that had printing all over them. He attempted to introduce us to a queer, slippery kind of fruit, which he called "banana," but had to give it up for the time being. After the meal, he had better luck with a curious piece of furniture on runners, which he called "rocking chair."

The children tried to conquer the chair as a cowboy tames a bronco, but they slipped and tumbled and could hardly do it.

> *One born and bred to the use of a rocking-chair cannot imagine how ludicrous people can make themselves attempting to use it for the first time.*

There was no bathtub in their apartment. "So in the evening of the first day my father conducted us to the public baths." When they came home it was evening and—this was amazing—the streets were bright.

> *So many lamps, and they burned until morning, my father said, and so people did not need to carry lanterns. In America, then, everything was free, as we had heard in Russia. Light was free; the streets were as bright as a synagogue on a holy day. Music was free....we had been serenaded, to our gaping delight, by a brass band of many pieces, soon after our installation on Union Place.*
>
> *Education was free. That subject my father had written about repeatedly, as comprising his chief hope for us children, the essence of American*

One immigrant wrote that "Castle Garden [above, where early immigrants arrived in New York] is a large building, a Gehenna, through which all Jewish arrivals must pass to be cleansed before they are considered worthy of breathing freely of the air of the land of the almighty dollar."

Jewish immigrants enter New York in 1882. It was a scary step to take. Some called Ellis Island "island of tears."

A family's New York City tenement. Some rooms never saw the sun or felt fresh air. Jacob Riis talked to a girl whose family could not afford to move to the front of their building. "They have the sun in there," she said. "When the door is opened the light comes right in your face."

opportunity, the treasure that no thief could touch, not even misfortune or poverty. It was the one thing that he was able to promise us when he sent for us; surer, safer than bread or shelter. On our second day I was thrilled with the realization of what this freedom of education meant. A little girl from across the alley came and offered to conduct us to school. My father was out, but we five between us had a few words of English by this time. We knew the world "school." We understood. This child, who had never seen us till yesterday, who could not pronounce our names, who was much better dressed than we, was able to offer us the freedom of the schools of Boston!...The doors stood open for every one of us.

It was May, almost the end of the school year, so the Antin children had to wait until September to begin school.

That day I must always remember, even if I live to be so old I cannot tell my name. To most people their first day of school is a memorable occasion. In my case the importance of the day was a hundred times magnified, on account of the years I had waited, the road I had come, and the conscious ambitions I entertained.

Father himself conducted us to school. He would not have delegated that mission to the President of the United States. He had awaited the day with impatience equal to mine, and the vision he saw as he hurried us over the sun-flecked pavements transcended all my dreams....The boasted freedom of the New World meant to him far more than the right to reside, travel, and work wherever he pleased; it meant the freedom to speak his thoughts, to throw off the shackles of superstition, to test his own fate, unhindered by political or religious tyranny.

In 1879, in Santa Fe, New Mexico, Flora Spiegelberger (who was Jewish), persuaded a Miss Carpenter (who was Presbyterian) to come and teach at a new school for girls. This would be the first nonsectarian school in Santa Fe. That meant that any girl, of any religion, could attend. Mrs. Spiegelberger adopted and raised an Indian son and daughter along with her own two daughters. When she needed advice on educational matters, she turned to her good friend Bishop Juan Bautista Lamy (who was Roman Catholic).

Mary Antin became the best student in her elementary school. When she wrote a poem about George Washington and it was published in a Boston newspaper, the whole school was proud of her. When she grew up she wrote her autobiography and called it *The Promised Land*. In it she wrote of her adopted nation's priceless heritage: it was the freedom and opportunity that let even the poorest immigrant—like herself—become rich in learning.

Mary Antin went to Boston Latin School, which still exists. These children are starting their day at New York City's Mott Street Industrial School in about 1892, when students first began pledging allegiance to the flag. How many stars did the flag have on it that year?

29 100 Candles

Happy birthday! One Centennial exhibit was a kindergarten, where children played and had lessons while a crowd watched.

President John Adams liked to tell the story of a Revolutionary War soldier who was heard crying out, "Attention, Universe, Kingdoms of the Earth, to the right about March!!" That soldier thought the world needed to pay attention to what was happening in this newly forming nation. He believed the kingdoms of the earth would not be the same again. He was right.

The nation's founders knew they were doing something never done before. They were making a nation based on self-government. They called it the "consent of the governed." Back then—at our birth in 1776, when the delegates to the Second Continental Congress signed the Declaration of Independence—people had little time to celebrate. They had to figure out how to fight Great Britain and win that independence.

But now it is 1876 and things are different. The experiment seems to be working. The nation has fought a terrible civil war—and become stronger for it. Americans are proud of themselves. The country is thriving. So everyone is excited by plans for a national birthday party; it will last for six months and be called the Centennial Exposition. This exposition is being held in a big park near Philadelphia; 200 buildings are needed to hold all the displays and activities.

Let's climb into a time capsule and join the party. In 1776 Philadelphia was tense; now it is mighty excited. America has never seen anything like this exposition. Philadelphia is all decked out in red, white, and blue. WELCOME TO THE CENTENNIAL, says a huge banner.

Ten million people will come to the Centennial before it closes. An article in the *Chicago Tribune* says: "Come at all events, if you have to live for six months on bread and water to make up for the expense." People come from all 38 states, and from foreign countries too. Most

A *centennial* is a hundredth anniversary.
An *exposition* is an exhibition, often a big, public one.

146

are astounded by what they see.

Machinery Hall, the most popular building, is 13 acres big. And that is small compared to the main exhibition hall—said to be the world's largest building—which covers 35 acres. The Woman's Building is almost as large as a football field. The very idea of a building showing women's accomplishments is an indication of changes that may be ahead. Philadelphia's Mrs. E. D. Gillespie—who happens to be Benjamin Franklin's great-grand-daughter—has been one of the most effective workers in promoting this Centennial. She has spoken before Congress and personally raised more money, through women's organizations, than any state (except New York and Pennsylvania). "The women of the whole country," says Mrs. Gillespie, "were working not only from patriotic motives, but with the hope that through this Exhibition their own abilities would be recognized and their works carried beyond needles and threads."

The Woman's Building is filled with women's inventions and artistry (and some needle-and-threadwork). One woman demonstrates a woman-designed darning machine, another shows a life-preserving mattress. A newspaper, the *New Century for Women*, is printed while visitors watch. But the most popular exhibit in the Woman's Building is a sculpture of a beautiful girl done in butter.

For a nickel you can ride around the Centennial grounds on a small railroad. What do you see? George Washington's coat, vest, and pants. A tropical garden under the glass roof of Horticulture Hall. The 40-foot-long arm and torch of the Statue of Liberty perched on top of

Top, the crowd on the Centennial's opening day. A Japanese observer said, "The first day crowds come like sheep, run here, run there, run everywhere. One man start, 1,000 follow....All rush, push, tear, shout ...say damn great many times, get very tired, and go home." Below, Machinery Hall's German cannons.

147

The exhibits included Lady Liberty's arm and torch (more on her in Book 8 of *A History of US*); tall glass tubes showing the quality and depth of the soils of Iowa; and "that glass case full of frozen fishes which as they reposed in their comfortable boxes of snow …did certainly appeal to some of the most vindictive passions of our nature;…during the hot months it will be cruelty to let them remain."

a souvenir stand. Real-looking artificial teeth. A gorgeous white hearse. Fancy coaches. Locomotives. Ben Franklin's printing press. A liberty bell made from tobacco. And the Capitol dome built of apples.

Iowa has sent samples of its soil as well as its farm produce. From Norway there are magnificent furs; from Egypt, saddles and mummies; and from Germany, the biggest and deadliest steel guns ever built.

The Centennial buildings have towers, domes, statues, banners, and flags. Some glass buildings seem just like crystal palaces. A fountain holds a marble sculpture of an Indian girl named Minnehaha, bronze dolphins, a lion's head, and marble basins to catch the splashing water. It is beautiful.

Are you hungry? You can eat a fancy meal or fill up on hot dogs, lemonade, hot roast potatoes, corn on the cob, popcorn, and ice cream. But don't eat too much—you might fall asleep, and you won't want to miss the carnival rides, or the toy displays, or the inventions.

People line up to see the new inventions. There are typewriters, hun-

"The Corliss engine," said writer William Dean Howells, "does not lend itself to description; its personal acquaintance must be sought by those who would understand its vast and silent grandeur....In the midst of this ineffably strong mechanism is a chair where the engineer sits reading his newspaper....Now and then he lays down his paper and clambers up one of the stairways...and touches some irritated spot on the giant's body with a drop of oil."

dreds of sewing machines, a newspaper press that prints and folds 15,000 news sheets in an hour, wallpaper printing machines, washing machines, and even machines to make machines. The technology is so amazing that some people say there is nothing left to invent.

The biggest hit of all is George Henry Corliss's colossal steam engine—the world's largest machine. It looms 40 feet high in Machinery Hall. Steam from the Corliss engine turns wheels; those wheels pull belts strung overhead; and the belts make 8,000 smaller machines work. President Grant sets the big steam engine going.

Just about everyone who enters Machinery Hall lines up to try a new gadget called a *telephone*. They say you can talk into the telephone and actually be heard in another room. When the emperor of Brazil visits the Centennial he puts his ear to it and cries out, "My God, it speaks." The *New York Tribune* calls it a "curious device," and asks, "Of what use is such an invention?" The newspaper answers its own question, "Some lover might wish to pop the question into the ear of a lady and hear for himself her reply, though miles away; it is not for us to guess how courtships will be carried on in the 20th century."

Shall we tell these people what telephones will soon do? Better not. Let them find out for themselves.

The telephone is Alexander Graham Bell's invention. Bell, a Scots immigrant, now a professor at Boston University, was working with the hard of hearing and the deaf when he built a device that let people see speech in the form of sound-wave vibrations. That made Bell believe that sound waves could be turned into electrical current and then back again into sound waves. He was right. But he had to study electricity and conduct experiments to make it happen. He tried many experiments. Finally—just two months before this Centennial opened—he did it. He made the sound waves of his voice travel over a wire.

Too bad Ben Franklin isn't here. He would be fascinated by the telephone and all the new devices on display. America is producing practical scientists: inventors who can turn ideas into products that make the world easier to manage. We are patenting inventions at an amazing rate.

The people around you would be astonished if you told them that the marvels here are just the beginnings of an age of invention. Soon there will be electric lights, record players, moving pictures, cameras, automobiles, airplanes, and more—much more. Most visitors at the Centennial will focus on the inventions and the products they can see and touch. But a few will think of other things, too. They will ask themselves, "Besides inventions and material things, what has America achieved in its first hundred years?" Have you thought about it? What has this nation actually accomplished?

The Emperor of Brazil hears a voice in Alexander Graham Bell's telephone. Bell's patent, no. 174,465, became the most valuable ever issued by the U.S. Patent office. In 1876 Western Union had a chance to buy the patent for $100,000. It seemed like too much money. Two years later the company couldn't buy the patent for $250,000—and was kicking itself.

30 How Were Things in 1876?

Uncle Sam welcomes representatives from many nations: "You see that we have grown a good deal in a century but just think how big we will be…in 1976."

Some people in the 19th century describe the rich and the poor as "two nations." In New York City, one boy works in the garment district while another celebrates his birthday.

Everyone enjoyed the party, but how about the serious questions? What had the United States accomplished in its first 100 years? How were things going?

Pretty well, thank you.

We were a free country with a constitution all the world envied. We had survived a terrible civil war and ended the horror of slavery. We had grown from 2.5 million citizens in 1776 to 46 million in 1876. Our exports (goods sent out of the nation) were greater than our imports (goods brought from other lands). That was a big change from the days when most manufactured things came from England. We were becoming a world industrial power. Still, our nation had problems, large problems, especially with that idea of fairness for everyone.

The Constitution now called for real equality for all. The 14th Amendment said *all persons born or naturalized in the United States…are citizens.* That meant all persons of every race, color, and religion. It further said:

At William Vanderbilt's house-warming fancy-dress ball in 1883 (some of the guests are shown below), Mrs. Cornelius Vanderbilt (above) came as "Electric Light," wearing white satin trimmed with diamonds and a diamond headdress.

No State shall make or enforce any law which shall abridge the privileges or immunities of citizens of the United States; nor shall any State deprive any person of life, liberty, or property, without due process of law.

It was a remarkable amendment. Nowhere else in the world did people have written guarantees like that.

Those written guarantees are making a difference, but good will is needed, too. In this year of 1876, in the South, congressional Reconstruction is almost finished, and, with it, most of the serious attempts to bring equality to blacks. The South is about to go back to state-supported inequality. It will pay for that stupidity by stagnating. That means it will not grow as fast economically as the rest of the country. But it is not only in the South that some people are treated unfairly; it is a national disgrace.

1876 is a big year. Out West the Sioux Indians win a victory at the battle of the Little Bighorn. It will turn out to be their last big win. For the American Indians the battle is almost over. They have lost control of most of their land.

On the Fourth of July of this year, American's leading women reformers, Elizabeth Cady Stanton and Susan B. Anthony, read the Declaration of the Rights of Women at the Centennial. Many people—some women as well as men—laugh at them.

Cornelius Vanderbilt, a railroad tycoon and one of the world's richest men, dies in 1876. Vanderbilt's Fifth Avenue mansion is fancier than most kings' palaces. His death makes people think hard about one of the problems the nation is facing: the distance between rich and poor, which is growing wider and wider. Jobs are hard to find. In New York City, in 1876, 900 people starve to death. The Children's Aid Society provides shelter for 11,000 homeless boys; 3,000 babies are left—abandoned—on its doorsteps.

While some people starve, others have so much money they do nothing much but show off. At Newport, Rhode Island, Cornelius Vanderbilt's grandson William lives in a sumptuous summer "cottage" with 70 rooms. At one of his parties, each guest is given a silver bucket and shovel to dig for rubies and diamonds buried in a sandbox in the middle of the dining-room table. Another wealthy family puts on the dog and invites 100 bow-wow guests to a party. The menu includes liver and rice, fricassee of bones, and dog biscuits. But instead of barking "thank you," the not-very-polite guests get into a big dogfight.

Let's hope that James Hazen Hyde's friends had better manners, because he threw a party that is said to have cost him $200,000. The average American factory worker, in 1876, earns $500—for a year's work. Twenty percent of American boys (and 10 percent of girls), ages 10 to 15, are working in factories or fields.

That's not the whole picture, though. The middle class—those people not rich or poor—is growing large in America, and living better than any middle class in all of history. Most middle-class people own their own homes, and are beginning to have leisure time to spend doing whatever they wish. Some are using it to play or watch those new games: baseball and football.

So, even though some Americans feel the country belongs to the rich and powerful, most are optimistic and confident. They believe in progress. They feel their problems can be solved. And they are eager to do what Thomas Jefferson suggested: pursue happiness.

In 1869, Rutgers University beat Princeton in a soccer match that is said to be the first intercollege athletic contest. Five years later, Harvard played McGill (in Toronto, Canada) at rugby and tied. Football came out of those two games. Rugby's scrum became football's scrimmage. By 1882, there were downs and measured yards. Yale had a famous coach, Walter Camp, who wrote football rule books, named "All-American" teams, and did a lot to define the modern sport.

William K. Vanderbilt (above) spent $2 million building his Newport house and $7 million decorating it. Partying was often a serious business for the Gilded Age, but these laurel-wreathed fellows seem to be heartily amused.

31 The Wizard of Electricity

Why do so many people never amount to anything? Because they don't think. It's astonishing what an effort it seems to be for many people to put their brains definitely and systematically to work. They seem to insist on somebody else doing their thinking for them. The individual who doesn't make up his mind to cultivate the habit of thinking misses the greatest pleasure in life. He not only misses the greatest pleasure, but he cannot make the most of himself.

—THOMAS EDISON

Thomas Alva Edison as a young man. He was world famous by the time he was in his thirties.

Sometimes disadvantages can be turned into advantages. Thomas Alva Edison had two big disadvantages: he had almost no schooling, and an accident made him deaf about the time of his 12th birthday.

Edison spent only a few months in school. His schoolteacher mother taught him some at home, but mostly he went to the library, got books, and taught himself. That ability to do things for himself helped him become the most successful inventor the world has ever known. Being deaf—or almost deaf—may have helped too. It allowed him to concentrate, without being distracted by conversation or sounds.

Thomas Edison was a lonely boy. He had a favorite science book and

Edison listens to his "talking machine" at 5 A.M. on June 16, 1888. He had just worked on it for three straight days to get it functioning properly.

154

a chemistry set, and he tried experiments over and over. In addition, he was a tinkerer: he liked to see how things worked. He would take something apart and then put it back together. Usually when he put it back together he made it work better.

He was fascinated by the new telegraph machines. When he was 11 he made one of his own. He intended to send messages back and forth between his house and another. But he needed to find a source of power, and he needed money for equipment and experiments; his parents had none to give him. So Edison went to work on a train, selling candy and newspapers. Since he had extra time on the long ride, the train people let him put his chemistry set in the baggage car. That baggage car was his first laboratory.

One warm summer day Tom Edison was standing on the train platform while his train took on water and fuel; he noticed a small child toddling across the tracks. Then he saw a freight train coming down those tracks. Edison jumped from the platform, dashed in front of the train, and rescued the child.

That action changed his life. The baby's father was a telegraph operator. He wanted to reward the brave 15-year-old candy seller. What could he do for him? "Teach me to be a telegraph operator," said Edison.

He soon learned Morse code and to send it rapidly. It was a useful skill. A good telegraph operator could work almost anywhere in the country. Edison got a job in Michigan; then he went to Boston, and then to New York. But he was soon bored just sending messages. His tinkering mind was at work; he thought he could improve the telegraph. So he invented a writing telegraph—a machine that wrote words, not just dots and dashes. He had other ideas, too. Western Union—the company that owned the telegraph system—asked him to sell his ideas. Edison figured they were worth about $3,000, but he told Western Union to set a price. When they offered $30,000, he was astounded! From that time on, Edison was a full-time inventor.

In 1876—that centennial year—he set up a laboratory in Menlo Park, New Jersey, with a team of gifted assistants. It was the world's first modern research laboratory. He called it an "invention factory." For the next five years he patented a new invention almost every month. He invented a motion-picture camera, and a projector; he built the first motion-picture studio (which was the beginning of the modern film industry). He invented a mimeograph (a type of copying machine), the storage battery, an electric locomotive, waxed paper, and composition brick. He discovered the movement of electrons in a vacuum—and that led to radio and modern electronics. He patented more than 1,000 inventions before he died.

Above, Edison's second wife, Mina. At right is his daughter Marion, nicknamed "Dot" in the telegraph's honor. (His son Thomas was—yes!—"Dash.")

In 1872 Edison invented a kind of electric typewriter that became the "writing" telegraph; later he found a way to send four messages over a telegraph line at the same time.

Above, Edison and his assistants beneath their light bulbs in the workroom at Menlo Park. When he first announced his intention to produce a cheap bulb in six weeks, the price of gas company shares crashed on the stock exchange. At right, a diagram of the carbon-filament light bulb.

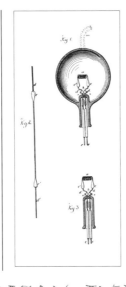

Between 1820 and 1910, America—and the world—was transformed by American inventions.

One day Edison left a drawing on the bench of one of his assistants. At the bottom of the paper was a note. "Make this," it said. "What will it do?" asked the assistant. "It will talk back," said Edison. It did. The phonograph came from that drawing. (And that led to the tape and CD players that you listen to.) Edison's first phonograph machine had to be cranked by hand, and the first record was of a song that Edison sang himself: it was "Mary Had a Little Lamb."

The most famous of all Edison's inventions was the electric light bulb. He didn't actually invent electric light. Others had been working on the problem. In 1859, Moses Farmer, in Salem, Massachusetts, had made two lamps with platinum filaments. Platinum is more expensive than gold. There was no way most people could afford light from that kind of bulb.

Edison announced he would take six weeks to develop a light bulb that could be made inexpensively. Well, he was wrong. It took him a little more than a year.

He knew that oxygen helps things burn, and he wanted his light to glow, not burn. The air would have to be pumped out of the bulb. And the glass should be clear, so the light would shine through.

SOME AMERICAN INVENTIONS 1830 to 1910

McCORMICK REAPER

STEEL PLOW

THRESHER

TELEGRAPH

SEWING MACHINE

SAFETY PIN

SKYSCRAPER

OIL WELL DRILLED

THE TYPEWRITER

AIR BRAKES (RR)

BARBED WIRE

1830 1840 1850 1860 1870

Edison hired a skilled glassblower to perfect a pear-shaped vacuum bulb.

Now he had to find a way for a *filament*—some kind of fiber, or something—to carry electricity, give off light, and last. He tried metals, papers, cork, and lemon peel. He plucked a hair from a friend's beard. He tested 6,000 vegetable fibers. Most of them fizzled or collapsed.

Finally he used a simple cotton thread made stiff with carbon. On October 21, 1879, Edison took some of that thread, put it in a glass bulb, pumped out the air, and turned on the current. The bulb "glowed like the setting sun in the dusk of early autumn," wrote one of his assistants. Edison and his staff sat down to wait. How long would the bulb glow?

They waited. And waited. Night came. In the morning the lamp was still glowing. The next night came. The bulb was burning as brightly as ever. That light shone for 45 hours. Edison had found what he sought; improving the bulb would be easy.

Edison was a practical genius. He had devised a lamp that would not only provide light but could be used by everyone. "We will make the light bulb so cheap," he said, "only the rich will be able to burn candles." But light bulbs are no good without sockets to plug them into, and light switches, and safety devices, and meters, and electric wiring, and dynamos to provide power. Edison invented some of those things and improved others. He turned the switch on the modern world of electricity.

Edison couldn't do that by himself. He needed help. He needed money—a lot of money. Would you lend money to a rumpled young

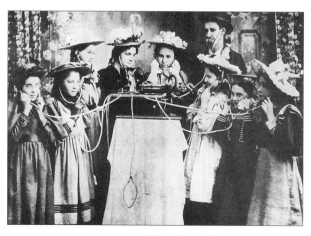

Listening to the phonograph in its early days was a group activity. As you can see, headphones are nothing new.

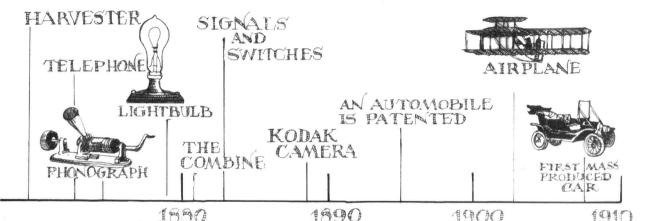

HARVESTER

TELEPHONE

LIGHTBULB

PHONOGRAPH

SIGNALS AND SWITCHES

THE COMBINE

KODAK CAMERA

AN AUTOMOBILE IS PATENTED

AIRPLANE

FIRST MASS PRODUCED CAR

1880   1890   1900   1910

Charles Brush put in the first electric light on Broadway (above) in 1881. Edison lit up the rest of New York City two years later.

man who said he could light a city? Grosvenor Porter Lowrey did. Lowrey was a Wall Street lawyer who played the violin and was fascinated by science. He went to J. P. Morgan, to Western Union, and to members of his own law firm. He asked them to finance Edison's schemes. A banker wrote, "The problem of lighting by electricity was far from being solved. Mr. Lowrey's…important contribution came from his ability to understand the scientific principles involved, and to explain them to his friends, thereby persuading them to invest large sums of money to underwrite Edison."

There was another problem: it was political. How was electricity to be taken from house to house? Who would give Edison the right to dig up city streets and lay wires?

On a cold December afternoon, Grosvenor Lowrey invited two groups of people to Menlo Park to visit Thomas Edison's laboratory. One group was made up of Wall Street bankers. They were all Republicans.

The other was a group of New York City politicians. (Many were Tammany Hall cronies of the now-dead Boss Tweed.) They were all Democrats. Group A and Group B didn't much like each other. But on that December day they did some pretending.

They all took a ferry across the Hudson River to New Jersey. A specially chartered train was waiting to carry them to Menlo Park. Lowrey explained that Edison wanted to electrify New York City. That meant laying wires under the ground. It meant building huge generators. It was an untried idea. It would be very expensive.

The bankers would have to contribute an enormous amount of money. The politicians would have to arrange for city rights. By the time the train reached Menlo Park, darkness had fallen. Thomas Edison greeted his visitors with

In 1908 crowds poured in to see an "Edison moving picture" in Litchfield, Minnesota. At the time it didn't matter much what the picture showed—it was enough that it moved at all.

a kerosene lamp in his hand. He took them around his laboratory. They could hardly see. One man stumbled on some wires on the floor. Some of the others began getting grumpy. Then Edison's assistant flipped a switch. What happened seemed incredible to the men. The laboratory was as bright as day! Edison took the men to the window. Lights burned outside; it was bright there, too.

Soon afterward the New York City Board of Aldermen passed an ordinance letting Edison install an electric lighting system. A New York paper noted, "For the first time in at least 40 years the politicians of this city have given away something for nothing. Could it possibly be a trend?"

On September 4, 1882, Thomas Edison stood in the offices of Morgan's New York banking firm. He pulled a switch. A few blocks away, six giant dynamos began to hum. Minutes later electric lights lit the bankers' offices and much of the rest of New York City. And, at almost the very moment the lights went on, a bolt of lightning cracked through New York's skies. Were the heavens angry that people were learning to control electricity?

Edison's inventive mind was soon hard at work on other projects. "Genius," said Thomas Alva Edison, "is 99 percent perspiration and one percent inspiration." Edison must have done a lot of perspiring. He slept only a few hours a day and expected those in his laboratory to keep up with him. He was so intent on his work he often forgot everything else. The day he married Mary Stilwell (it was Christmas Day), he went to his lab to do some work and quite forgot his bride. He worked far into the night.

He was at the first theater performance lit by electricity. When the lights began to flicker he dashed into the cellar, took off his fancy clothes, and shoveled coal to keep the steam boiler supplying electricity. He was still shoveling during a banquet held upstairs in his honor.

Edison was our most gifted and famous inventor—but, at the end of the 19th century, America seemed a land of inventors. In 1815, the United States Patent Office gave patents to 173 inventions. Between 1860 and 1890, the number was 440,400.

Edison was never concerned with "pure science"—he wanted to make things that worked. He said about his son Theodore: "His forte is mathematics. I am a little afraid...he may go flying off into the clouds with that fellow Einstein."

J. P. Morgan was a very rich banker. (His story is told in Book 8 of *A History of US*.) Morgan's house was one of the first in the world to be lighted by electricity. But when the wealthy Vanderbilts put electricity in their house, two wires crossed and a fire started. They tore all the electrical wiring out of their mansion and went back to candles and oil lamps. Morgan and the other financiers reaped great profits from electricity.

32 Jim Crow— What a Fool!

A white man who blacked his face made Jim Crow's song and dance known all over the world.

Jim Crow wasn't real. He was the name of a character in a song—a song about a black man who sang and danced and never gave anyone any trouble.

> *Wheel about, turn about, dance jest so —*
> *Every time I wheel about I shout Jim Crow.*

The Jim Crow character was well known on the stages of the North. Later, offstage, his name came to stand for an evil policy—the policy of separation of the races, also known as segregation.

Before the Civil War, the North had something the South didn't have. It had segregation. In the North, the races were separated—not by law, but by habit. Usually blacks (and Indians and Asians) were not welcome in white hotels, restaurants, schools, or theaters. They could not get good jobs. The name for that policy of segregation became *Jim Crow.* Jim—that old fool dancer—sang and danced and acted as if he was content with things as they were, but he was crying on the inside.

In the antebellum South there was

No Forty Acres, and No Mule

The former slaves expected to be given free land with their freedom. There was a rumor, and it was widely believed, that they would each get "forty acres and a mule." After all, African-Americans had worked for nothing for generations. Some congressmen—the Radical Republicans— tried to see that black people got land. But it didn't happen. Blacks got freedom—and poverty. Frederick Douglass, the abolitionist leader who had been a slave, wrote in 1882 that the slave "was free from the individual master, but the slave of society. He had neither money, property, nor friends. He was free from the old plantation, but he had nothing but the dusty road under his feet."

slavery but not segregation. Whites were the masters, but whites and blacks often lived and worked together. (This is a little tricky; be sure you understand it.) In the South blacks and whites often had close ties. Although it may seem strange to us now, some slaves and slave owners liked each other.

After the Civil War, things stayed much the same in the North. Segregation by habit continued. In the South nothing was the same. Remember, Reconstruction was a time of confusion, change, and experiment.

Here's a quick review of what happened to the newly freed men and women during Reconstruction. At first the southern states passed terrible *black codes* that practically made black people slaves again. Then Congress sent army troops south; some northern abolitionists went with them. Years of congressional Reconstruction followed: many blacks got a chance to go to school, to vote, and to hold public office.

Some white Southerners didn't seem to want a fair interracial society. Some did. Most were confused. For blacks, Reconstruction began as a time of hope. But conditions were terrible in the South. The war had left almost everyone poor.

In 1877, when congressional Reconstruction ended and the army troops left, the South was on its own. Once again, white Southerners turned to the wrong leaders. These mean-spirited leaders blamed blacks and Northern carpetbaggers for all the troubles of the South.

The Redeemers (those who wanted to end Reconstruction) began by counting votes incorrectly. That's called *voter fraud*. They paid some poor blacks for their votes. Eventually they took the *franchise*—the vote—from blacks and poor whites by passing poll taxes and other laws meant to keep blacks from voting. Without the vote, blacks were powerless.

Once the army had left, Southerners were on their own again—and soon, for many blacks, "freedom" was indeed worse than slavery.

Jim Crow began dancing across the southern land. Lawmakers passed laws that made it a crime for the races to be together. Soon blacks and whites had to ride in separate railroad cars, go to separate schools, get buried in separate cemeteries, pray in separate churches, and eat in separate eating places.

There was nothing blacks could do about it. Remember, they were no longer able to vote.

Some white Southerners tried to stop the Jim Crow laws. In 1897, an editorial in a Charleston, South Carolina, newspaper said:

The poll tax (later found to be unconstitutional) meant you had to pay a tax to vote. Other laws made people pass a written test in order to vote. Sometimes different tests were given to blacks and whites.

CIVIL RIGHTS AT WALLACK'S THEATRE.

During Reconstruction, well-dressed blacks could go almost anywhere. (White or black, poor families like the Georgia farmers at right—however respectable—were not welcome in fancy places.) But as Jim Crow took hold, scenes of discrimination like the one above, where the manager suddenly finds all the seats are taken, became more and more common.

Jim Crow's Beginnings

Jim Crow was born because of a belief in "white supremacy." Many whites—North and South—believed that whites were better than other peoples. It was an old idea—a racist idea. It's tempting to say, "I'm better than everyone else." It's tempting, but boastful, unkind, and untrue. Racism, throughout history, has always led to evil action. In the 20th century, in Germany, racism brought about the Holocaust—the murder of 6 million Jews.

Most 19th-century white Americans—North and South—were good people. Some of them had a mistaken belief and some didn't even know it was wrong. They didn't think about it. Not thinking can sometimes be as evil as bad thinking.

The common sense and proper arrangement, in our opinion, is to provide first-class cars for first-class passengers, white and colored....To speak plainly, we need, as everybody knows, separate cars or apartments for rowdy or drunken white passengers far more than Jim Crow cars for colored passengers.

But common sense had fled from the South. By the time the new century arrived, Jim Crow was going wild. Black citizens, who had high hopes after the Civil War, were now sometimes worse off than before. Hundreds of blacks were lynched—hanged—by white mobs, and police did nothing about it. The 14th Amendment says:

No State shall make or enforce any law which shall abridge the privileges or immunities of citizens of the United States; nor shall any State deprive any person of life, liberty, or property, without due process of law; nor deny to any person within its jurisdiction the equal protection of the laws.

What does that mean, "the equal protection of the laws?" Read the amendment again. Make sure you understand your rights. The 15th Amendment says:

> *The right of citizens of the United States to vote shall not be denied or abridged by the United States or by any State on account of race, color, or previous condition of servitude.*

Were the Jim Crow laws constitutional or unconstitutional? What do you think? Were the southern states defying the Constitution when they passed laws and taxes that prevented blacks from voting?

Of course they were. In 1896 some Louisiana citizens went to the Supreme Court to see what they could do about it. Six years earlier, the Louisiana General Assembly had passed a bill that said railroad companies must "provide separate but equal accommodations for the white and colored races" on passenger trains.

Homer Plessy's great-grandmother was African. Everyone else in his family had a European background. Plessy's skin was white. But, according to the racists, anyone with any African blood at all was black. So Homer Plessy was considered a black. His friends in New Orleans wanted to show the ridiculousness of the whole idea of racial categories. That's why they chose white-skinned Homer Plessy for a test case.

Reconstruction was a hopeful experiment that died. Jim Crow, here illustrating the song about him, was a poor fool who lived too long.

They got Plessy to sit in the white section of a railroad car. When the conductor was told that he was black, Plessy was arrested, charged with breaking the law, and put in jail. Plessy and his attorneys said that the "separate but equal" law was unconstitutional. New Orleans's Judge John H. Ferguson said they were wrong. Plessy's case made it all the way to the Supreme Court. It was called *Plessy* v. *Ferguson*. *Plessy* v. *Ferguson* is famous for being one of the worst decisions the Supreme Court ever made. It changed the lives of millions of people—and not for the better.

The court agreed with Judge Ferguson! After that, Jim Crow really went wild. He danced and sang like fury. Separate-but-equal became the way of the South. Before the *Plessy* case there were many examples of integration in the southern states. Twenty years later there were almost none.

Sometimes the Supreme Court makes mistakes (as in the *Dred Scott* and *Plessy* decisions), but it seems to correct those mistakes over time.

The promise of hope (above) that many saw in the 1875 Civil Rights Act vanished with the *Plessy* case, whose decision, read by Justice Henry B. Brown (below left) upheld segregation—despite the disagreement of Justice John M. Harlan (right).

Frederick Douglass, speaking at the World's Fair in Chicago in 1893 said, "Men talk of the Negro problem, there is no Negro problem. The problem is whether American people have loyalty enough, honor enough, patriotism enough, to live up to their own Constitution....we Negroes love our country. We fought for it. We ask only that we be treated as well as those who fought against it."

One Supreme Court justice, John Marshall Harlan, disagreed with the other justices. He said so in strong language. He wrote a dissenting opinion in *Plessy* v. *Ferguson*. Justice Harlan said:

> *In view of the Constitution, in the eye of the law, there is in this country no superior, dominant ruling class of citizens. There is no caste here. Our Constitution is colorblind, and neither knows nor tolerates classes among citizens.*

Justice Harlan's words are worth remembering. Every lawyer knows them. Every American should. It would take time, but, finally, the Supreme Court agreed with Harlan. Separate is *not* equal. In 1954, the Supreme Court changed its mind. *Plessy* v. *Ferguson* was reversed. Segregation was found to be unconstitutional. Jim Crow was kicked off the stage.

A Law Review for the Constitution

As you know, we have a three-branch government. The executive branch (the president) proposes laws, the legislative branch (Congress) passes laws, and the judicial branch (the courts) may be called on to review laws to see if they are constitutional or not. That process—called judicial review—began in America with Chief Justice John Marshall and a case called *Marbury* v. *Madison*. Judicial review has now spread to many nations. (In some countries, including Britain, legislatures pass laws and also decide if they are constitutional. There is no review by judges there.)

Many thinking people believe that judicial review is one of the great safeguards of our liberty. (Some don't. They say it gives too much power to a small group of unelected justices.) Be sure you understand how judicial review works. The Supreme Court does not look at all the laws Congress passes and check them over to see if they are constitutional or not. If Congress passes a law—no matter what it is—it is the law. In order for the Supreme Court to decide if that law is constitutional or not, someone must bring a case to it. That means someone must be willing to go through the long process of appealing the case from court to court. He or she must be willing to risk losing. It is a process that has involved many Americans—rich and poor, of every race, color, and religion.

33 Ida B. Wells

Ida B. Wells married and had three children, but even with a household to run, she never stopped working to get her protests heard.

Some people were murdering other people (yes, it was murder), and nearly everyone—and that means the president, Congress, state leaders, and most ordinary people—was looking the other way. They didn't want to face what was happening.

But Ida B. Wells was a newspaper woman and she had no intention of looking aside. So she wrote about the murders—which were called lynchings. What happened to Wells? Her newspaper's printing press was smashed. People said they would kill her, and they meant it. She moved, bought a gun, and kept writing.

Not many people wanted to read what she had to say. Many were embarrassed by the murders. Maybe they thought they would go away. Maybe they thought Ida Wells would go away.

For a while, she did. She went to England, where she gave speeches about the lynchings. People there listened. They were shocked. Stories and articles and editorials were written about her in important English newspapers. Lords and ladies and some of Parliament's leaders met Ida B. Wells. Meetings were held to protest the murders. A few people in the United States were upset by the fuss in England. But most people just didn't seem to know what to do, so the lynchings continued.

But I need to let Ida B. Wells tell you her own story:

> *I was born in Holly Springs, Mississippi, before the close of the Civil War. My parents, who had been slaves and married as such, were married again after freedom came. My father had been taught*

Ida Wells couldn't call 911—but you can.

Freedom's Journal was the country's first newspaper published and edited by black people.

A black writer traveling in the South in 1870, when Ida Wells was a little girl, wrote: "While I am in favor of universal suffrage, yet I know that the colored man needs something more than a vote in his hand; he needs to know the value of a home life."

The Freedmen's Association was run by O. O. Howard. Howard University in Washington, D.C., was named for him.

the carpenter's trade, and my mother was a famous cook....My father [called Jim] was the son of his master...and one of his slave women, Peggy....He was never whipped or put on the auction block, and he knew little of the cruelties of slavery....My mother...was born in Virginia and was one of ten children. She and two sisters were sold to slave traders when young, and were taken to Mississippi and sold again....her father was half Indian....She often wrote back to somewhere in Virginia trying to get track of her people, but she was never successful.

Holly Springs was a small town with the kind of people—black and white—who, after the war, attempted to live together in harmony.

Our job was to go to school and learn all we could. The Freedmen's Aid had established one of its schools in our town....My father was one of the trustees and my mother went along to school with us until she learned to read the Bible. After that she visited school *regularly to see how we were getting along.*

Ida's parents were community leaders. Blacks and whites in Holly Springs rode streetcars and trains together. It was Reconstruction and, mostly, there was racial harmony.

But when voting time came, Ida's father, Jim Wells, was told to vote for the Democratic candidates. Wells voted Republican. His boss wouldn't let him back in the carpenter's shop where he worked. Ida's dad didn't hesitate. He bought some tools, rented a shop across the street, and went into business for himself. It was a lesson Ida would remember.

Ida was the oldest of seven children, and a good student. She planned to go to college. One summer, when she was in high school, she went to visit her grandparents on their farm. Yellow fever was raging through the South. The epidemic was said to have started when a ship docked in Norfolk, Virginia, with some sick people. It traveled like a summer tornado. The germs had no prejudice. They killed men, women, and children of every size and color.

The epidemic was worst in the cities, so people fled from cities to small towns. Some of those people came to Holly Springs. They brought their germs with them. Soon people in that town were sick. Jim and Lizzie Wells helped some of the sick people. And they prayed with them.

One day three horsemen rode up to the farmhouse gate and handed Ida a letter. This is what it said:

Jim and Lizzie Wells have both died of the fever. They died within twenty-four hours of each other. The children are all at home and the Howard Association has put a woman there to take charge of them. Send word to Ida.

Ida Wells was 16, and her parents were dead.

If she went home she was sure to get the fever. Everyone told her to stay with her grandparents. Besides, there were no passenger trains running. No one was traveling while the fever raged. She went anyway. She went on a freight train. It was draped with black cloth to honor two conductors who had died of the fever. The train engineer told her to stay away

Yellow fever killed thousands before it was discovered that mosquitoes carried the disease and that they could be controlled.

His Name Was Mudd!

During the yellow-fever epidemic, Dr. Samuel Mudd worked for weeks—day and night—caring for sick people. That took courage—some doctors had died of the fever, and others didn't want to go near the epidemic victims. Mudd was no ordinary doctor; he was a prisoner on the Dry Tortugas off the Florida coast—convicted of conspiracy to kill the president!—and everyone knew his name. Mudd had set the broken leg of Abraham Lincoln's assassin, John Wilkes Booth. Was he part of a conspiracy? Or just a doctor doing his job? Most of the evidence suggests that he was innocent. After Mudd's selfless doctoring during the epidemic (all of his patients lived), President Andrew Johnson pardoned him.

Before the Civil War there was a saying "the *mud* press" to describe newspapers that threw mud at people's reputations. A person with mud on his name had a poor reputation. But after Dr. Mudd's trial the expression "his name is mud" became common.

from Holly Springs. She asked him why he was driving the train; after all, he could get the fever, too. He said, "Someone has to do it."

"That's exactly why I am going home," said Ida. "I am the oldest of seven living children. There's nobody but me to look after them now. Don't you think I should do my duty, too?"

SOUTHERN HORRORS.

LYNCH LAW

IN ALL

ITS PHASES

Miss IDA B. WELLS.

Price, · · · Fifteen Cents.

THE NEW YORK AGE PRINT.
1892.

Chicago's congressmen and Ida Wells called on President William McKinley: "We refuse to believe this country, so powerful to defend its citizens abroad, is unable to protect its citizens at home," said Wells.

Ida Wells always put duty ahead of fear or personal safety. She always spoke up, which is what she did when she got home. But the first thing she learned made her weep some more: her baby brother was dead. Now there were six children. Friends had made plans for them. The youngest children were to be split up between different families. A handicapped sister was to be put in an institution. Ida was thought old enough to take care of herself.

> When all this had been arranged…I, who had said nothing before and had not even been consulted, calmly announced that they were not going to put any of the children anywhere; I said that it would make my father and mother turn over in their graves to know their children had been scattered like that and that we owned the house and…I would take care of them.
>
> I took the examination for a country school-teacher and had my dresses lengthened, and I got a school six miles out in the country. I was to be paid the munificent sum of twenty-five dollars a month.

She rode a mule the six miles to school. For Ida Wells, childhood was over.

But I began this chapter by talking about murder and trips to England. How did Wells get from Holly Springs, Mississippi, to London, England?

With her pen. Ida B. Wells became a famous journalist. She began by sending letters to some black newspapers.

As you probably gathered, Wells was a no-nonsense person who said exactly what she thought. Her letters made sense. She told the truth. The letters became popular. Several newspapers wanted to carry them. Soon she had a regular newspaper column.

Black newspaperwomen were nothing new. Back in 1850, before the Civil War, Mary Ann Shadd Cary founded the *Provincial Freeman*, a weekly read by black refugees in Canada. Cary was called "one of the best editors in the Province."

After the war a number of women were writing for black newspapers, but none of them wrote as well as Ida B. Wells. Thomas Fortune, a leading newsman, wrote of her:

> *She has become famous as one of the few of our women who handles a goose quill…as handily as any of us men. She is rather girlish looking in physique with sharp regular features, penetrating eyes, firm set thin lips and a sweet voice.…She has plenty of nerve, she is smart as a steel trap, and she has no sympathy for humbug.*

The *Washington Bee* described her as a "talented young school-marm…about four and a half feet high." (She was actually taller than that, but she wasn't tall.)

But writing did not pay enough to allow Wells to quit teaching. One day she boarded a train to Memphis. She was going to teach there. Ida sat in the ladies' coach, as she had always done. But the South was changing. The conductor wouldn't take her ticket. He said blacks had to sit in the smoking car. Ida Wells wouldn't budge.

> *As I was in the ladies' car I proposed to stay. He tried to drag me out of the seat, but the moment he caught hold of my arm I fastened my teeth in the back of his hand.*
>
> *I had braced my feet against the seat in front and was holding on to the back, and as he had already been badly bitten he didn't try it again by himself.*

The conductor got two men to help him. They tore her dress and dragged her from the train.

Wells hired a lawyer and sued the railroad. The judge said she was right and awarded her $500. But the Chesapeake and Ohio Railroad appealed the case to the Tennessee Supreme Court. That court reversed the lower court's decision. Ida B. Wells had to pay a fine. In her diary she wrote:

> *I had hoped such great things from my suit for my people generally. I have firmly believed all along that the law was on our side and would when we appealed to it, give us justice. I feel shorn of that belief and utterly discouraged.*

She was 22.

Understanding?

Isaiah Montgomery

By 1890, Isaiah Montgomery was the only black left in the Mississippi legislature. (Isaiah was a brother of Mary Virginia Montgomery and the founder of the farming community at Mound Bayou, Mississippi.) That year Mississippi passed an "Understanding Law." It said voters had to "understand" the Constitution before they could vote. Officials asked blacks very difficult questions about the Constitution; if they could not answer, they could not vote. Whites usually got asked easy questions. Isaiah Montgomery voted for that unfair law. He didn't want to upset the whites; he thought he would have more influence that way. Ida Wells was outraged, and said so in her newspaper, the *Free Speech*.

Mr. Montgomery came to Memphis to explain, but although we never agreed that his course had been the right one, we became the best of friends, and he helped to increase the circulation of the paper wonderfully.

34 Lynching Means Killing by a Mob

Ida B. Wells followed up her articles on the Memphis lynchings of 1892 with a short book called *Southern Horrors*.

Ida B. Wells had many friends in Memphis, but Thomas Moss and his wife were special friends. Moss was a postman who saved his money and, along with two partners, opened a grocery store. Their store was across the street from a white-owned grocery. The white grocer didn't like having competition. He threatened the new grocers. Then he and his friends marched on the store—with guns. Moss and his friends had guns, too. Three white men were wounded.

The three black grocers were taken to jail. But they weren't safe there. A mob invaded the prison, took the three men, and filled their bodies with bullets. It was a *lynching*. It was murder by a mob. The dictionary says lynching is "to execute without due process of law."

> *I have no power to describe the feeling of horror that possessed members of the race in Memphis when the truth dawned upon us that the protection of the law was no longer ours.*

The 14th Amendment guarantees all Americans due process of law. In other words, every American has a right to a fair trial. But the

Anti-lynching laws were finally passed in the mid-20th century.

Vigilantes took the law into their own hands in the West, too.

READ AND REFLECT!

Murders High-handed Outrages

THIS IS TO NOTIFY ALL WHOM IT MAY CONCERN!

WILL BE SUMMARILY DEALT WITH!

AND PUNISHED AS OF OLD!

VIGILANCE COMMITTEE.

Charles Russell, an artist whose paintings of cowboys and the West became famous, sketched a lynching party on its way home after the deadly deed was done.

Constitution is only as good as the citizens who enforce it.

Between 1882 and 1930, 4,761 people were lynched in the United States. It happened in the North and West as well as the South. Only the New England states were without lynchings. Whites and Asians were lynched. But most lynchings were in the South, and most victims were black. In Mississippi, 545 people were murdered; in Texas, 492; in Louisiana, 388; Montana had 93 lynchings (only two of those victims were black).

In the West it was called *vigilante* (vih-juh-LAN-tee) justice, but it was never just, because in our nation real justice must come through the legal process. If people take the law into their own hands—and decide who is guilty and who is not, without a fair trial—there is no point to laws and government. Then you have *anarchy*, or rule by no one. What you don't have is fairness. Most people thought the lynch mobs were poor and uneducated people. That was a myth. In every section of the country, the mob murderers—the lynchers—had the consent and often active participation of community leaders.

Most citizens—black and white—didn't want to hear about the murderers. That was because of another myth. That myth said that the victims of the mobs were all criminal men who were attacking women or doing something else that was wrong. (Murder is never right, but if these men were really bad, maybe they were just getting what they de-

Mob Rule

Lynching gets its name from Charles Lynch of Virginia. During the Revolutionary War, Lynch organized posses to rid the area of Tories. They terrorized and killed. After the war, the Virginia General Assembly excused Lynch's actions as war measures, saying, "measures taken…may not be strictly warranted by law, although justifiable from the imminence of the danger." Lynch's actions, and those of vigilante squads in the West, have been called "establishment violence." Are there times when people should take the law into their own hands? What are the alternatives?

Bus boycotts would happen again in the 20th century, when a civil-rights movement finally ended segregation in the South.

served. That was what many people believed. At first, even Ida Wells believed that.)

But Wells knew Tom Moss. She knew he was a fine citizen. He hadn't attacked any women. Could it be that some of the others who had been lynched were also innocent victims? Ida Wells decided to find out. What she found was that only one third of the lynch victims had even been accused of attacking women. Some women and children had been lynched. Some of the dead were victims of mistaken identity. Wells wrote about what she found in the *Free Speech*. She said other newspapers—the big, fancy ones—weren't writing the truth. Ida Wells was fired from her teaching job.

When the city of Memphis refused to even try to find Tom Moss's murderer, she told her readers to leave Memphis. "There is only one thing we can do—leave a town which will neither protect our lives and property, nor give us a fair trial." Within two months, 6,000 black people had left Memphis. Then she organized a boycott: she told her readers to stop riding the streetcars. When blacks stopped riding the streetcars, white businesses began to suffer.

> The superintendent and the treasurer of the City Railway Company came into the office of the Free Speech *and asked us to use our influence with the colored people to get them to ride on the streetcars again. When I asked why they [said] the colored people had been their best patrons.: "So your own job then depends on Negro patronage?" I asked. And although their faces flushed over the question they made no direct reply.*

In her next article in the *Free Speech*, Wells "told the people to keep up the good work."

But Ida Wells was soon in danger herself. After one of her articles appeared, the *Free Speech* was attacked and its presses ruined. It happened while Ida Wells was on her way to New York. Editor Thomas Fortune met her at the ferry. "We've been a long time getting you to New York," he said. "But now I'm afraid you'll have to stay." Some people in Memphis were talking about lynching Ida B. Wells. It was 30 years before she could go south again.

Of course, she didn't stop fighting for justice. Ida Wells moved to

Chicago, married lawyer Ferdinand Barnett, raised a family, gave speeches, went to England, founded a community center, worked for women's suffrage—and kept writing. Finally, when others were ready to do something about lynching, they could turn to Ida B. Wells's articles and find well-researched data and truths that were hard to face.

Lynchings were not often spontaneous. Most were carefully planned. The crowd here set up the beam, tied the rope around the man's neck, and threw him out the window.

173

35 A Man and His Times

Booker T. Washington was a man of his times: they were Jim Crow times; they were lynching times.

I was born a slave on a plantation in Franklin County, Virginia. I am not quite sure of the exact place or exact date of my birth, but at any rate I suspect I must have been born somewhere and at some time.

So begins the autobiography of Booker T. Washington, who was born five years before the Civil War began. By the end of the 19th century, he was one of the best-known men—black or white—in America. Here is more of the autobiography:

My life had its beginning in the midst of the most miserable, desolate, and discouraging surroundings. This was so, however, not because my owners were especially cruel, for they were not, as compared with many others. I was born in a typical log cabin, about fourteen by sixteen feet square. In this cabin I lived with my mother and a brother and sister till after the Civil War, when we were all declared free.

The little cabin where Booker lived was also the plantation kitchen. His mother was the plantation cook. Cooking was done in an open fireplace. The family slept on the dirt floor, on bundles of rags.

I was asked not long ago to tell something about the sports and pastimes that I engaged in during my youth. Until that question was asked it had never occurred to me that there was no period of my life that was devoted to play. From the time that I can remember anything, almost every day of my life has been occupied in some kind of labor; though I think I would now be a more useful man if I had had time for sports.

Booker T. Washington became a useful man anyway. He was 10 when the Civil War ended and his mother moved her small family to West Virginia. Booker went to work in a salt furnace. But his mother was determined that he get an education. He wanted to go to school, too.

I had no schooling whatever while I was a slave, though I remember on several occasions I went as far as the schoolhouse door with one of my young mistresses to carry her books. The picture of several dozen boys and girls in a schoolroom engaged in study made a deep impression upon me, and I had the feeling that to get into a schoolhouse and study in this way would be about the same as getting into paradise.

Getting into paradise wasn't easy. Booker's family needed the money he earned. He watched other black children going to school, and he ached to go, too. Finally he found a way. He got up early, worked until school began, went to school, and then went back to work.

Then someone told Booker Washington about a Negro college in Hampton, Virginia. He decided that was where he wanted to go. He didn't know anyone there, or if they would accept him; he just headed east until he got to Hampton. It was 500 miles. He arrived without any money and got a job as a janitor to pay for his studies.

Hampton Institute provided vocational training for blacks. That means it taught students to be farmers, carpenters, teachers, brickmakers, or to do other useful jobs. Students learned skills and to take pride in their work.

After the Civil War, in the South, few people wanted to work hard. Many blacks thought hard work was slaves' work—and they didn't want to be reminded of slavery. Whites didn't want to work hard either—they had always had slaves to do that. So no one valued hard work.

The teachers at the institute (who were mostly Northern whites) understood that people who work hard get things done and feel good about their accomplishments. At Hampton, teachers worked right along with students. Booker T. Washington was one of the best of those students.

So when the president of Hampton Institute was asked to recommend someone to head a new training institute for blacks at Tuskegee,

Tuskegee students putting up buildings to form part of their school. This achieved three goals at once. The school got built; the students learned important and useful trades; and their labor paid for their tuition.

Now, in regard to what I have said about the relations of the two races, there should be no unmanly cowering or stooping to satisfy unreasonable whims of Southern white men, but it is charity and wisdom to keep in mind the 200 years' schooling in prejudice against the Negro which the ex-slaveholders are called upon to conquer.

—BOOKER T. WASHINGTON

Do you think that statement is wise or foolish?

Agricultural Revolutionary

Here is a personal confession: I have fallen in love with George Washington Carver! Not only was he one of America's great scientists, he was also sweet-natured and determined to make his life count. So I'm a bit annoyed that there isn't room for a chapter about him; but this book is long enough. Maybe it's for the best—now you'll have to find out about him on your own.

Carver sparked an agricultural revolution in the American South, urging farmers to change from soil-exhausting crops like cotton and tobacco to those, such as peanuts and sweet potatoes, that fed people and the soil, too. He couldn't bear to waste things, and that led him to develop more than 400 synthetic materials from common crops and agricultural leftovers (cheese, dyes, even synthetic marble!). When Booker T. Washington invited him to Tuskegee, Carver found a home, adding prestige to that university. The Missouri plantation where he was born to a slave mother was declared a national monument in 1953.

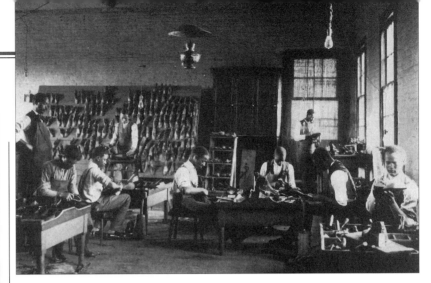

Students at Tuskegee learned practical accomplishments like shoemaking, but they also learned sophisticated mechanical skills, such as electrical engineering; Tuskegee's chapel was the first building in the county to be wired for electricity.

Alabama, he suggested Booker T. Washington for the job.

Aside from that recommendation, the people at Tuskegee didn't know much about him. They were expecting a white man to head the school. Washington was expecting a "building and all the necessary apparatus ready for me to begin teaching." They were both surprised.

What Washington found at Tuskegee, in June of 1881, was one building and an old church. This is how he described the schoolhouse:

> This building was in such poor repair that, whenever it rained, one of the older students would very kindly leave his lessons and hold an umbrella over me while I heard the recitations of the others.

Washington rolled up his sleeves and went to work. He and his students cut down trees, cleared land, dug wells, and built buildings. By 1900, Tuskegee had 40 buildings and some fine teachers. Booker T. Washington was renowned as the voice of the black people.

He was quite a speaker. A newspaperman described him as "a remarkable figure; tall, bony, straight as a Sioux chief, high forehead, straight nose, heavy jaws, and strong, determined mouth, with big white teeth, piercing eyes and a commanding manner."

He spoke so well that he often left audiences cheering. Contributions poured into Tuskegee Institute. The school's graduates went out and trained others. Booker Washington said that blacks must first gain economic freedom by learning working skills and getting good jobs; then they could battle for other kinds of freedom. Washington became a national hero.

But some blacks weren't happy with just economic freedom. They were American citizens. They wanted to vote, to ride on buses with everyone else, and to go to the same schools. They wanted to send lynchers to jail and kill Jim Crow. One man began to criticize Booker T. Washington's style of leadership, and that shocked a lot of people.

36 A Man Ahead of His Times

W. E. B. DuBois as a junior at Fisk University. He became the first black person to receive a Ph.D from Harvard.

History is full of surprises. When you're living through it, there is no predicting how it will turn out.

Consider those two leaders Booker T. Washington and W. E. B. DuBois (do-BOYS). By the end of the century, everyone had heard of Booker T. Washington. He was said to be the best-known southern man since Jefferson Davis. He was much loved by both races. When Booker Washington spoke, people listened. President Theodore Roosevelt invited him to dinner. Wealthy men like Andrew Carnegie and John D. Rockefeller gave money to the causes he supported.

Now the average person—black or white—had never even heard of W. E. B. DuBois. Those who had knew that he was brilliant, that he was an African-American thinker and writer, and that he had earned a Ph.D. from Harvard. They also knew that his ideas often got him in trouble. They knew that he had been fired from some jobs and that he had money problems. But what seemed shocking was that he often disagreed with the great Booker T. Washington, and said so in writing.

How could anyone disagree with Booker Washington? All the newspapers agreed with him; so did most college professors.

Booker T. Washington (right) was a wonderful speaker, but in his speeches he urged blacks to work hard and not agitate for civil rights.

177

I awoke at eight and took coffee and oranges, read letters, thought of my parents, sang, cried &c (O yes—the night before I heard Schubert's beautiful unfinished symphony...). Then I wandered up to the reading room, then to the art gallery, then to a fine dinner....Then went to Potsdam for coffee and saw a pretty girl. Then...took a wander, supped on cocoa, wine, oranges and cake, wrote my yearbook and letters— and now I go to bed after one of the happiest days of my life.

Night—grand and wonderful. I am glad I am living....I wonder what the world is—I wonder if life is worth the striving. I do not know—perhaps I never shall know—but this I do know: be the Truth what it may I shall seek it on the pure assumption that it is worth seeking—and Heaven nor Hell, God nor Devil shall turn me from my purpose, till I die.

—W. E. B. DuBois,
WRITTEN ON HIS 25TH BIRTHDAY
WHEN HE WAS A STUDENT IN BERLIN

"We refuse to surrender...leadership...to cowards and trucklers," wrote DuBois. "We are men; we will be treated as men."

Why, even the *president* believed his ideas were right and sound.

Well, as I said, history is full of surprises. Today historians say that Booker T. Washington was a man of his times who, without meaning it, may have held his people back. DuBois, they say, was a man who had important things to say to whites as well as blacks—even though, at the time, few people seemed to be listening.

Why do some historians think Booker T. Washington may have held his people back? What did he do wrong?

He compromised. Sometimes compromise is the best thing to do. Booker T. Washington thought he was doing right, and maybe he was. What do you think?

Remember, he lived in the days of Jim Crow. He lived in the days when two or three blacks were lynched— murdered—each week. What did he do about that? Almost nothing.

Washington compromised with the whites who were in power. He told the white leaders that blacks wanted jobs, that all they wanted was a chance to earn money. He believed that if blacks had economic opportunities, other opportunities would follow. So he headed a school that trained blacks in working skills. Now that was fine. He taught the value of hard work, and that was important. But he also told white audiences that if blacks could have jobs and economic opportunities they wouldn't demand social equality. They would live with Jim Crow. That made him very popular with some whites. He didn't try to change things. He didn't fuss. Many people are afraid of fuss and change.

During the years that Washington was the leading American black, Jim Crow grew mightier and mightier. He also grew sillier and sillier. In Mississippi, Jim Crow segregated soda machines, and in Oklahoma he segregated phone booths!

But W. E. B. DuBois wouldn't compromise with anyone. He wanted full equality. Nothing less would do. Besides,

DuBois understood that in a democracy all citizens must be treated fairly. Prejudice is not democratic. He wasn't the only one who felt that way. It was the theme of America's Declaration of Independence. Carl Schurz said, "If you want to be free, there is but one way, it is to guarantee an equally full measure of liberty to all your neighbors." By denying full freedom to blacks—and women and Asians—America was weakening its democratic government. W. E. B. DuBois understood that.

He worked to bring the vote to women, he spoke out against anti-Semitism (which is prejudice against Jews), he worked to get fair treatment for immigrant groups, he tried to lynch Jim Crow. He couldn't do it, but his words and ideas helped those who would do it after him. Some people say he was the father of the civil rights movement of the 20th century.

William Edward Burghardt DuBois was born in Great Barrington, Massachusetts. His family had lived there for a long time. One ancestor fought in the Revolutionary War. The DuBois family was respected.

Each spring, in Great Barrington, all the townspeople gathered for a town meeting where they decided how to run the schools and how to spend the town's money. While he was still a boy, DuBois went to the town meetings and learned that democracy was "listening to the other man's opinion and then voting your own, honestly and intelligently."

But it was during a visit to his proud grandfather—who would never agree to any form of segregation—that he learned about his African heritage. He met other young African-Americans and admired their physical beauty, their enthusiasm, and their good minds. He wondered, could he be both American and black? Or must he choose? Here is his answer to that question:

> *We are Americans, not only by birth and by citizenship, but by our political ideals.*

DuBois understood that each American has a double inheritance that includes:

In 1905, DuBois (second from right, middle row) and a group of other black Americans met at Niagara Falls. They announced that they were working for the exercise of all civil rights without regard to color.

The Negro folk-song—the rhythmic cry of the slave—stands today... as the most beautiful expression of human experience born this side of the seas.
—W. E. B. DuBois, *THE SOULS OF BLACK FOLK*

"Liberty trains for liberty," said W. E. B. DuBois. "Responsibility is the first step in responsibility." What did he mean?

"It seems to me," said Booker T.,
"It shows a mighty lot of cheek
To study chemistry and Greek
When Mister Charlie needs a hand
To hoe the cotton on his land,
And when Miss Ann looks for a cook,
Why stick your nose inside a book?"

"I don't agree," said W. E. B.
"If I should have the drive to seek
Knowledge of chemistry or Greek,
I'll do it. Charles and Miss can look
Another place for hand or cook.
Some men rejoice in skill of hand,
And some in cultivating land,
But there are others who maintain
The right to cultivate the brain."

"It seems to me," said Booker T.,
"That all you folks have missed
  the boat
Who shout about the right to vote,
And spend vain days and sleepless
  nights
In uproar over civil rights.
Just keep your mouths shut, do not
  grouse,
But work, and save and buy a house."

"I don't agree," said W. E. B.,
"For what can property avail
If dignity and justice fail?
Unless you help to make the laws,
They'll steal your house with
  trumped-up clause.
A rope's as tight, a fire as hot,
No matter how much cash you've got.
Speak soft, and try your little plan,
But as for me, I'll be a man."

"It seems to me," said Booker T.—

"I don't agree," said W. E. B.

—Dudley Randalll

• the responsibilities and rights that go with American citizenship

• and the personal richness of his or her own ethnic roots.

DuBois believed that our diverse backgrounds enrich us all. No people originated on this continent, not even the Native Americans. It is the collection of heritages that make America special.

W. E. B. DuBois's ideas would help lead blacks—and all people—to pride in their achievements. His own accomplishments were intellectual. DuBois wrote many books and edited several magazines. He was one of the founders of the NAACP—the National Association for the Advancement of Colored People—an organization of blacks and whites formed to fight racial injustice. "The problem of the 20th century is the problem of the color line," said DuBois. He didn't think color should be a problem.

DuBois had good ideas, but he didn't have much talent for people. He didn't know how to work well with others. So most people didn't pay attention to him.

Booker T. Washington's talent was in healing differences; in helping people get along. Washington said there was a time when he hated whites, but he quit when he realized that "hating the white man did him no harm and…was narrowing up my soul and making me a good bit less of a human being." Think about those words. Are you sometimes tempted to hate others? What does that do for you?

As it turned out, the ideas and example of both Booker T. Washington and W. E. B. DuBois are important to all Americans. Each wanted to make our country do what it was always meant to do—be fair to all people. As DuBois said:

We are Americans, not only by birth and by citizenship, but by our political ideals….And the greatest of those ideals is that ALL MEN ARE CREATED EQUAL.

180

37 End Words

"The mass of mankind has not been born with saddles on their back," said Jefferson, "nor a favored few booted and spurred, ready to ride them."

Back in 1776, when Thomas Jefferson, John Adams, and Benjamin Franklin talked about forming a new nation, they knew they needed to find just the right words to make it special. There had to be a reason for the colonies to break away from England. They believed there were good reasons.

They thought about the Puritans, who came to America to build "a city on a hill"—a place people could look up to—a community that would inspire others.

It was surprising how many others—like Roger Williams, William Penn, and James Oglethorpe—had come to do the same thing. None of them found it easy to build a really good sociey. Besides, they didn't even agree on what made a good society. Jefferson, Adams, and Franklin understood that. But it didn't discourage them. They were determined to start the new nation on a noble path, which is exactly what they did.

Thomas Jefferson, who sat down to write out their ideas, came up with great words. He said:

> *all men are created equal, that they are endowed by their creator with certain unalienable rights, that among these are life, liberty, and the pursuit of happiness.*

If ever anyone asks you if words are important, just quote those words from the Declaration of Independence. The world has not been the same since they were written.

Would these citizens—the Tucker brothers of Dayton, Ohio—be given a chance at life, liberty, and the pursuit of happiness?

Jakob Mithelstadt and family were Germans who migrated to Russia before moving to America. They arrived in New York City in 1905 in search of new life and possibilities.

Jefferson got US off to an extraordinary start. But it was just a start.

Ideas build on each other. Andrew Jackson came along and said, "Let the people rule," and democratic government exploded forth. Abraham Lincoln talked of government "of the people, by the people, and for the people," and democracy was defined.

Government for the people. You'd think that would be easy. It isn't. Because most people keep yelling, "For *me*, for *me*," and those with the loudest voices drown out the weak and powerless.

So government has to have keen ears. It needs to hear those who can't always make themselves heard: children, the handicapped, the poor.

Idea building is a process that goes on and on. Elizabeth Cady Stanton and her friends at Seneca Falls, New York, pointed out a flaw in Jefferson's words. So they changed them to *all men and women are created equal.* Susan B. Anthony was willing to stand trial because she believed women should be able to vote.

When Thaddeus Stevens had himself buried in an integrated cemetery he knew exactly what he was doing. But in case others didn't get it, he had the words *Equality of man before his creator* chiseled on his tombstone. He was adding another layer to the promise of America.

Nothing could stop Ida B. Wells in her fight for justice—neither unfairness nor threats. Her words and her example would inspire laws to protect Americans of every color.

A little girl named Mary Antin was brought to America because it offered religious freedom. In Russia, where her family had lived, Jews were persecuted.

The people in this demonstration demanding votes for women range from society ladies to sweatshop workers—and include 600 men.

Many came to the land where church and state were respectfully separated. Jews like Irving Berlin, George Gershwin, Aaron Copland, and Leonard Bernstein would soon be composing American music. Others would become scientists, doctors, artists, writers, and teachers. They would love their new home.

So would Germans like Carl Schurz, who wrote:

> *If you want to be free, there is but one way. It is to guarantee an equally full measure of liberty to all your neighbors.*

Some newcomers came to find freedom; some came for opportunity. One young boy wrote to his friends back in Sweden that in America his cap was not

> *worn out from lifting it in the presence of gentlemen. There is no class distinction between high and low, rich and poor, no make-believe, no "title-sickness."...Everybody lives in peace and prosperity.*

America wasn't perfect, but it certainly was heading in the direction of freedom, equality, and happiness for all its citizens.

Which was exactly what Jefferson, Adams, and Franklin had in mind, back in 1776, when they got together to write a declaration.

The Gesell family came from Germany in search of opportunity and settled in Alma, Wisconsin. Gerhard Gesell (whose son Arnold became a famous child psychologist) snapped his relatives during a big family reunion.

183

Chronology of Events

1865: Andrew Johnson becomes 17th president when Abraham Lincoln is assassinated

1865: Johnson outlines Reconstruction; pardons former Confederates swearing loyalty to U.S.

1865: the states ratify the 13th Amendment to the Constitution; slavery is abolished

1866: a Civil Rights Act guarantees citizenship and equal rights to black Americans

1866: Confederate veterans in Tennessee establish the racist and terrorist Ku Klux Klan society

1866: Jesse Chisholm organizes the first great cattle drive between Texas and Abilene, Kansas

1867: Nebraska becomes the 37th state

1867: U.S. buys Alaska from Russia for $7.2 million

1867: Oliver Kelley founds the Grange to help protect farmers' interests

1867: Christopher Sholes patents the first typewriter

Mar. 1868: Congress impeaches President Johnson; the Senate acquits him by one vote

July 1868: the states ratify the 14th Amendment, giving federal protection of individual rights

1868: Republican candidate General Ulysses S. Grant, Civil War victor, is elected 18th president

1869: the 15th Amendment gives black males the right to vote

1869: Susan B. Anthony founds National Woman Suffrage Association

1869: the transcontinental railroad is completed at Promontory Point, Utah

1869: The women of Wyoming Territory are the first in the U.S. and territories to get the vote

1870: Hiram Revels of Mississippi becomes the first black man elected to the U.S. Senate

1871: Congress repeals the Northwest Ordinance; Indian tribes are no longer independent nations

1871: Barnum & Bailey's circus opens in Brooklyn

1872: President Grant is tainted by the Crédit Mobilier scandal; despite this he is reelected

1872: Congress closes the Freedmen's Bureau

1872: Yellowstone becomes the first national park

1872: half a million immigrants enter the U.S.

1873: Joseph Glidden invents barbed wire

1875: Congress passes a Civil Rights act banning discrimination in public places

1876: the Centennial Exposition opens in Philadelphia

1876: Sioux and Cheyenne warriors led by Crazy Horse and Sitting Bull defeat General George Custer's cavalry at the Little Bighorn in Montana

1876: Rutherford B. Hayes becomes 19th president after a disputed election; he pledges to abolish martial law and end Reconstruction

1876: Alexander Graham Bell patents his invention of the telephone

1877: the Nez Perce Indians, led by Chief Joseph in their flight to Canada, surrender to U.S. troops

1877: Southern states introduce a poll tax and other devices aimed at undermining civil rights

1879: Belva Lockwood becomes first female lawyer admitted to practice before the Supreme Court

1879: Thomas Edison demonstrates his carbon-filament incandescent light bulb

1880: Cyrus McCormick's Chicago factory is producing over 100 reaping machines a day

1880: the census shows the U.S. with a population of over 50 million and New York the first city with a population over one million

1881: Booker T. Washington founds the Tuskegee Institute in Alabama

1882: Congress passes the Chinese Exclusion Act, forbidding Chinese to enter the U.S. for 10 years

1884: Mark Twain publishes *Adventures of Huckleberry Finn*

1886: Supreme Court overturns *Yick Wo* v. *Hopkins* as discriminatory and contrary to 14th Amendment

1892: Ida B. Wells publishes her exposé of the Memphis lynchings

1896: with the *Plessy* v. *Ferguson* decision, the Supreme Court rules that segregation by race is constitutional

More Books to Read

We would like to thank the children of the Malibu School in Virginia Beach, Virginia, for their help in compiling this list of books for further reading.

Pam Conrad, *Prairie Visions,* Harper Trophy, 1993. Solomon Butcher was a photographer who traveled around Custer County, Nebraska, more than a century ago, taking pictures of settlers, cowboys, and children, sod houses, horses, and windmills. He wrote down people's stories and jokes, too. This excellent book has many of his pictures and stories.

Mary Ann Fraser, *Ten Mile Day,* Henry Holt, 1993. This is a picture book—with clear writing, too—about how the workers of the Central Pacific Railroad managed to lay 10 miles of transcontinental track in one day when one of the railroad owners bet that it couldn't be done.

Russell Freedman, *Children of the Wild West,* Clarion, 1983. A good book, illustrated with many interesting photographs, that gives a lot of details about life in the West from the point of view of Plains Indians as well as pioneers.

Paul Goble, *Death of the Iron Horse,* Bradbury, 1987. Many stories have been made up about trains wrecked by Indians, but it actually happened only once, in 1867, when a group of Cheyennes derailed a Union Pacific freight train in Nebraska. This little book, with its beautiful drawings, tells the story.

Hamilton Holt, ed., *The Life Stories of Undistinguished Americans,* Routledge, 1990. These articles and interviews were published in a newspaper, the *Independent,* around 1900, and then put together in a book. The "undistinguished Americans" include a Polish sweatshop girl, a Greek peddler, an Illinois farmer's wife, a black farm worker, a Mohawk Indian, and a Japanese servant.

Patricia MacLachlan, *Sarah, Plain and Tall,* HarperCollins, 1988; *Skylark,* HarperCollins, 1994. Anna's mother died when Anna's brother Caleb was born. Their father puts an ad in the paper for a mail-order wife to come and look after them on the prairie; they get Sarah from Maine. She writes to them: *I will come by train. I will wear a yellow bonnet. I am plain and tall.* The children hope desperately that she will stay. Two short but very good books by a terrific writer.

David Willis McCullough, ed., *American Childhoods,* Little, Brown, 1987. Like *Undistinguished Americans,* this is an adult anthology, with great short extracts from longer books by real grownups remembering their lives as children—people like U. S. Grant, Charlotte Forten, Ohiyesa, and Mary Antin.

Cornelia Meigs, *Invincible Louisa,* Little, Brown, 1933. You probably know that Louisa M. Alcott wrote a famous children's book called *Little Women* (see Book 5 of *A History of US*); but did you know that Louisa struggled half her life to provide for her family because her impractical father could not manage, or that she worked as a nurse during the Civil War? This splendid biography is almost as exciting as a novel.

Milton Meltzer, *The Black Americans: A History in Their Own Words,* Harper Trophy, 1987. This excellent collection of documents includes the words of famous black Americans—Frederick Douglass, Robert B. Elliott, Ida B. Wells—and also of unknown former slaves, "exoduster" farmers, and army veterans.

Scott O'Dell and Elizabeth Hall, *Thunder Rolling in the Mountains,* Houghton Mifflin, 1992. Sound of Running Feet was the real daughter of Chief Joseph of the Nez Perce, and in this marvelous but sad novel she tells the story of the tribe's attempted flight to Canada, their eventual capture by U.S. cavalry, and their forced migration to reservations.

Mark Twain, *Adventures of Huckleberry Finn* (first published 1884; available in many editions). This is a great and funny book. Just read it.

Laura Ingalls Wilder, *The "Little House" books,* Harper & Row uniform edition, 1953. Laura Ingalls based these stories (*Little House in the Big Woods, Little House on the Prairie, The Long Winter,* etc.) on her own life as a pioneer child and young woman in Wisconsin, Kansas, and Dakota Territory. If you haven't read them, read them. If you've read only one or two, read the others.

Index

Picture Credits

In this fairly typical 19th-century view of the wild West, "a drunken cowboy undertakes to reform the rules of ballroom etiquette in Leadville, Colorado, and gets a setback."

Company, San Francisco; **59 (top):** Oakland Museum, Andrew J. Russell Collection; **59 (bottom):** Union Pacific Railroad; **60:** Southern Pacific Company, San Francisco; **61 (top):** Oakland Museum of Art; **62 (bottom left):** Lucius Beebe; **62 (bottom right):** Library of Congress; **63 (top):** Oakland Museum of Art; **64:** *Harper's Weekly,* 1858; **65:** *Harper's Weekly,* November 13, 1866; **66 (top):** *Harper's Weekly,* May 29, 1869; **66 (bottom),** **67 (top):** Library of Congress; **67 (bottom):** Lucius Beebe; **68 (top):** State Historical Society of Wisconsin; **68 (bottom):** Kansas State Historical Society; **69 (top):** Nebraska State Historical Society; **69 (bottom):** National Archives; **70-71:** Montana Historical Society, Helena; **73 (left):** gift of Charles Dalton, Copperas Cove, Texas; **73 (right):** Cather family; **76:** State Historical Society of Wisconsin; **77:** Bettmann Archive; **78:** State Historical Society of Wisconsin; **79:** Bettmann Archive; **80:** National Anthropological Archives, Smithsonian Institution; **81 (top):** Walters Art Gallery, Baltimore; **81 (bottom):** Milwaukee Public Museum; **82:** National Museum of Natural History; **83 (bottom):** National Archives; **84 (top):** State Historical Society of Colorado; **84 (bottom left, bottom right):** New York Public Library; **85:** National Archives; **86:** New York Public Library Picture Collection; **86–87:** Corcoran Gallery of Art, Washington, D. C.; **88:** Library of Congress; **89:** Montana Historical Society, Helena; **90:** Heye Foundation, Museum of the American Indian, New York; **91:** Washington State University; **92 (top):** National Archives; **92 (bottom), 94 (top left):** Smithsonian Institution; **94 (top right):** Montana Historical Society; **94 (bottom):** Bowdoin College Library; **95:** *Harper's Weekly,* October 21, 1871; **96 (top):** Benjamin Blom, Cityana Gallery; **96 (bottom):** *Harper's Weekly,* June 4, 1870; **97:** *New York Times,* August 19, 1871; **98 (top):** *Leslie's Illustrated,* February 19, 1870; **98 (middle, bottom):** *Scientific American,* March 5, 1870; **100:** *Harper's Weekly,* October 7, 1876; **101 (right):** New-York Historical Society; **102 (right):** Barnum Museum, Bridgeport, Connecticut; **103 (top):** Library of Congress; **103 (bottom):** James Dunwoody; **104 (bottom):** Albany Institute of History and Art, Albany, New York; **106 (top):** Mark Twain Papers, Mark Twain Memorial, Hartford, Connecticut; **106 (bottom):** *Scientific American,* May 24, 1844; **107 (top):** Library of Congress; **108 (left):** New-York Historical Society; **108 (right):** Library of Congress; **109 (bottom):** Chicago Historical Society; **110:** Mark Twain Memorial, Hartford, Connecticut; **111 (left):** New-York Historical Society; **111 (right):** New York Public Library; **112:** New York Public Library; **113:** Library of Congress; **116 (left):** National Portrait Gallery, Smithsonian Institution; **116 (right);** Museum of the City of New York; **118 (left):** New York Public Library; **119 (right):** Culver Pictures; **120 (top):** Augustus F. Sherman Collection, Ellis Island Immigration Museum; **120 (bottom):** South Dakota State Historical Society; **121 (right):** Library of Congress; **122 (bottom):** *Harper's Weekly,* April 1, 1882; **123 (left):** Idaho State Historical Society; **123 (right):** Mary and Weston Naef Collection of Stereographs; **124 (left):** Bancroft Library, University of California, Berkeley; **124 (right):** California Historical Society, San Francisco; **125 (bottom):** Library of Congress; **126 (top):** *Harper's Weekly,* March 25, 1872; **128 (left):** *Harper's Weekly,* May 20, 1882; **129:** courtesy John Carrey; **130:** Wyoming State Archives and History Department; **131 (top):** Western History Research Center, University of Wyoming; **131 (bottom):** Library of Congress; **132 (top):** Denver Public Library, History Department; **132 (bottom):** Wyoming Historical Society; **133 (left):** National Portrait Gallery, Smithsonian Institution; **133 (right):** Library of Congress; **134 (left):** University of Nebraska; **135 (left):** Library of Congress; **135 (middle):** National Portrait Gallery, Smithsonian Institution; **135 (right), 136:** Library of Congress; **139 (middle):** Curator's Office, U.S. Supreme Court; **140:** Library of Congress; **141 (top):** Israel Museum, Jerusalem; **142 (top):** collection of the artist, Lionel S. Reiss; **144 (top):** Library of Congress; **145 (bottom):** Museum of the City of New York; **146 (top):** Julian Wolff; **147, 148 (left):** Free Library of Philadelphia; **150:** National Geographic Society; **151 (top left):** Library of Congress; **151 (top right):** Lewis Hine Collection, George Eastman House; **151 (bottom):** Byron Collection, Museum of the City of New York; **152:** New-York Historical Society; **153:** Museum of the City of New York; **154 (top):** National Portrait Gallery, Smithsonian Institution; **154 (bottom):** Edison National Historical Site; **155 (bottom):** from Will Ray, *Art of Invention,* Pyne Press, Princeton, New Jersey; **155 (top left, top right), 156 (middle), 157 (top):** Edison National Historical Site; **158 (bottom):** Library of Congress; **159:** collection of Oliver Jensen; **161 (right):** *Harper's Weekly,* October 24, 1874; **162 (right):** University Archives, Bancroft Library, University of California, Berkeley; **163:** New-York Historical Society; **164 (top):** *Harper's Weekly,* April 24, 1875; **164 (bottom left, bottom right):** Library of Congress; **165 (left):** Schomburg Center for Research in Black Culture, New York Public Library; **165 (right), 166 (top):** New York Public Library Picture Collection; **167 (bottom):** American Museum of Natural History, Smithsonian Institution; **168 (top):** Johnson Collection, Library of Congress; **169:** Mississippi Department of Archives and History, Jackson; **170 (top):** University of Chicago Library, Special Collections; **171:** Hammer Galleries, New York; **172–173:** New York Public Library Picture Collection; **175, 176 (top):** Johnson Collection, Library of Congress; **176 (bottom):** Library of Congress; **177 (top):** New York Public Library Picture Collection; **177 (bottom):** Library of Congress; **178:** New York Public Library Picture Collection; **179:** from Langston Hughes and Milton Meltzer, *A Pictorial History of the Negro in America;* **180:** National Portrait Gallery, Smithsonian Institution; **181 (top):** Independence National Historical Park, Philadelphia; **181 (bottom):** Library of Congress; **182 (top):** Ellis Island Museum; **183, 191:** Library of Congress

A Note From the Author

Do you remember, before the Civil War, when Sojourner Truth spoke at a women's rights convention in Akron, Ohio?

"There were very few women in those days who dared to 'speak in meeting,'" wrote Frances Dana Gage, who was chairwoman at Akron. Some men (they were ministers) had taken control of the convention and were lecturing the women on how to behave—and, said Gage, "the boys in the galleries, and the sneerers among the pews, were hugely enjoying [themselves]."

When Sojourner Truth rose, many of the women were really fearful. They thought the situation would get even worse. "'Don't let her speak, Mrs. Gage, it will ruin us. Every newspaper in the land will have our cause mixed up with abolition…and we shall be utterly denounced.'" But chairwoman Gage "begged the audience to keep silence a few moments."

Then Sojourner Truth stunned her listeners with a powerful speech, in which she said, *I have ploughed, and planted, and gathered into barns, and no man could head me. And ain't I a woman?* and the whole nation learned of the power of her words and presence.

Each time I read that speech, I try to imagine that I'm at the convention—so I can feel the excitement in the hall. But I'd never thought to ask: Who was Frances Gage? And then I read a poem by Mrs. Gage (it was set to music and was a popular song), and I realized: she's the woman who let Sojourner Truth speak!

It's connections like that one that make history exciting. A lot of the people that you and I read about knew each other. They worked together, or were at odds with each other. Seeing the links between them helps you see a bigger picture.

If you start doing that, you'll discover that many not-so-well-known people are part of that big picture; often they make a real difference in their times. (And that means that any of us can do the same thing.)

Yes, ordinary people, the kind you meet every day, can be extraordinarily effective, and interesting too. (Wait until you read about the civil-rights movement in the 20th century and see what young Americans did.)

Frances Gage was one of those effective people. When she died—in 1884—Clara Barton (who founded the American Red Cross) wrote a letter about her to Lucy Stone (who was a famous feminist). They both called her Aunt Fanny (Fanny was a nickname for Frances). This is some of what Clara Barton said:

Dear, noble, precious Aunt Fanny, the heart so warm, the sympathies so quick and ready, the sensitive, shrinking modesty of self, the courage that scoffed at fear when the needs of others were pleaded; the friend of the bondman [slave] and oppressed, who knew no sect, sex, race or color, but toiled on for freedom and humanity.…if only her clarion voice could ring out.

Well, you can hear Frances Gage's clarion voice in her poem, titled "A Hundred Years Hence," which means 100 years from then—and that was almost 150 years ago!

One hundred years hence what a change will be made, In politics, morals, religion and trade…

…Our laws then will be noncompulsory rules, Our prisons converted to national schools:

Lying, cheating and fraud will be laid on the shelf, Men will neither get drunk nor be bound up in self, But all live together as neighbors and friends, Just as good people ought to a hundred years hence.

Then woman, man's partner, man's equal shall stand, While beauty and harmony govern the land…

Oppression and war will be heard of no more, Nor the blood of a slave have its print on our shore…

Instead of speech making to satisfy wrong, All will join the glad chorus to sing Freedom's song, And if the millennium is not a pretence We'll all be good brothers a hundred years hence.

When I first read those words I chuckled. "The poet was naive," I thought. But Frances Gage may have the last laugh; maybe her optimism isn't so far wrong. Slavery is finished, women are men's equal, and many of us do live as "good neighbors and friends." Yes, oppression and war are still heard of; too many of our citizens are in prisons; and we need laws, not "non-compulsory rules." We have a long way to go, but maybe we've come further than we realize.

Together, most of us do sing Freedom's song—we cherish a common belief in liberty and justice for all. So, like Aunt Fanny, let's believe in the millennium (look that word up in your dictionary!). Let's try to make the world a better place.